Conservation Agriculture

CONSERVATION AGRICULTURE

P. K. Sethunath
Nitish K. Tiwari

Conservation Agriculture

First Published : 2014

ISBN 978-81-8342-336-6

Published by :

CRESCENT PUBLISHING CORPORATION
4819/24, Mathur Lane,
Ansari Road, Darya Ganj,
New Delhi - 110002
Mob. : 9711991838
Fax : 91- 011- 23257835
e-mail : crescentbook@gmail.com

Printed at :
Roshan Offset Printers
Delhi

Preface

Conservation agriculture is a form of farming meant to ensure soil health, promote sustainable farming, and prevent the damaging overuse of land. Gaining popularity in the 21st century, conservation agriculture aims to help farmers keep land healthy and productive by maintaining sustainable growing methods. There are several strategies to ensuring the health and continued fertility of land through conservation agriculture, many of which are outlined in guidelines from the Food and Agricultural Organization of the United Nations.

One of the most important principles that forms the backbone of conservation agriculture is reducing soil damage by discouraging tilling. A second major factor in conservation agriculture is the planting of cover crops. Although the UN guidelines regarding conservation agriculture acknowledge that some application of pesticides may be necessary for commercial crops, it is also recommended that farmers minimize herbicide use to protect the natural biodiversity of the soil and local environment. By protecting soil and ensuring a healthy, thriving bio-system, many experts believe that crops are more likely to flourish.

This book aims to show how conservation agriculture can increase crop production while reducing erosion and reversing soil fertility decline and restoring the environment. It will serve as invaluable source of references to the scientists, students, teachers, farmers and extension officers.

Authors

Contents

1

Introduction

Concerns about soil erosion affecting soil productivity in rainfed areas have resulted in an emphasis on trying to stop negative effects on crop yields attributed to erosion and runoff. This has been attempted by putting cross-slope barriers in fields designed to catch or divert soil and water moving downslope. This approach has not been particularly successful either in halting the problems or in raising yields, resulting in disillusionment among farmers. Money has been spent to little effect and damage to land has not been stopped.

However, if the emphasis is shifted towards the soil as a habitat for roots and if soil loss and runoff are recognized as consequences of prior damage to soil porosity, a different perception emerges. This is based on more positive thinking which considers first the soil conditions that allow plant roots to function optimally, and then the improvements necessary to bring any current inadequate state of the root habitat to that desired condition. Land uses would ideally be cross-matched with variations in land suitability with respect to erosion hazard - i.e. the most protective forms of land use would be allocated to places with the highest hazards of erosion. However, especially for farmers with few resources and small farms, low yields of subsistence crops may dictate that they be planted on all land units irrespective of erosion hazard. In both situations however, improving soil conditions to meet the needs of plant roots will often greatly reduce problems of soil loss and runoff.

Key goals of improving and maintaining excellent soil conditions for and with roots include:

— increasing the reliability of plant production in the face of unpredictable variations in the weather and other hazards of the environment;

— reducing production costs and raising net returns to producers;

— increasing the quality of the land and its resilience to extreme weather conditions.

On increasingly large areas of Latin America there has been a revolution in agricultural practice over the past 30 years. The adoption of zero tillage methods of crop production by large numbers of farmers provides convincing validation of the value of such conservation-effective forms of agriculture, in agronomic, environmental, economic and social terms. This is being achieved on farms whose sizes range from less than twenty hectares to thousands of hectares and in a wide range of ecological zones.

Conservation agriculture (CA), as defined during the First World Congress on Conservation Agriculture (1-5 October 2001) promotes the infiltration of rainwater where it falls and its retention in the soil, as well as a more efficient use of soil water and nutrients leading to higher, more sustainable productivity. It also contributes positively to environmental conservation. In many environments conservation agriculture can be considered the ultimate soil and crop management system. Conservation agriculture has been successfully implemented in both small-scale and large-scale farming, where it has given economic benefits as well as improved water resources.

HISTORY

Zero tillage has been successfully practised in the United States for several decades, with regular annual growth in the total area. In Latin America there has been an impressive rate of adoption and accelerating growth over the past two decades.

Brazil and Paraguay suffer erosive rainstorms of very high intensities during the southern summer, which result in severe damage year after year. On almost all cultivated land, soil tillage for crop production, often with heavy disk ploughs followed by disk harrowing, resulted in many problems. These included:

— loss of the porous organic covering of the forest floor where land had been cleared;

— pulverization of surface soil together with compaction of the subtillage layer;

— loss of organic matter from the upper soil layers by rapid oxidation from the exposed surface;

— loss of potential soil moisture as runoff;

— reduction in soil depth by erosion of topsoil, resulting in losses of seeds and fertilizers, and causing additional replanting costs;

— declining flow and drying-up of streams and rivers during dry seasons.

Downstream there were problems with eroded sediments clogging urban water purification plants, sedimentation in stream valleys and reservoirs, damage to bridges and roads. A common response was to construct conservation banks on the contour, such as broad- and narrow-based bunds to control runoff and soil erosion. However, they did not stop erosion occurring on uncovered soil. Infiltration of runoff was impeded by the severe compaction along the channel where it collects. The channel beds are probably the most compact lines in the entire field.

As time passed and the runoff and erosion problems continued, larger and larger banks were built, but without conspicuous success in halting the problem. Declining productivity and profitability on family farms resulted in collapsing net farm incomes, falling land prices and families leaving their farms for some other livelihoods.

In 1972 there were 500 hectares under residue-based zero tillage on one farm in southern Brazil. The technique spread slowly at first, because of scepticism and insufficient knowledge. There was a lack of appropriate equipment, suitable cover crops and weed control techniques. As the economic and technical advantages of residue-based zero tillage became apparent the rate of spread accelerated, largely as a result of farmer-to-farmer contacts. By 2001 in Brazil, there were more than 13 million hectares managed in this way.

In the State of Santa Catarina, Brazil, residue-based zero tillage has been adopted on 400 000 ha by 1998-1999 within a programme to promote these systems. As a result, some or all of the improved practices were spontaneously adopted on a further 480 000 ha outside the formal remit of the project, from a base of 120 000 ha in 1993-1994. The State's small farmers have been ingenious in devising their own equipment and methodologies to fit zero tillage to their individual circumstances, together with governmental and non-governmental arrangements for their technical and institutional support.

In Paraguay, zero tillage was first used in the late 1970s but was not widely adopted on mechanized medium and large farms until 1990. It had expanded to 20 000 ha by 1993, to 250 000 ha by 1995-1996 and to 480 000 ha by 1997, which represents 51 percent of the total cultivated area of Paraguay.

Implementing Conservation Agriculture

Planting a crop in the residues of the previous crop, which is the essence of conservation agriculture, is fast becoming a successful and sustainable cropping practice, especially in the subhumid tropics. Implicit in this practice is the absence or limitation of tillage practices that incorporate surface residues or disrupts soil porosity.

The quantity of crop residues produced is clearly very important and varies greatly with crop type, variety and yield. Invariably there are residues of weeds associated with crop residues that also contribute to soil cover, especially during the initiation of no-till. Large quantities of crop residues are usually obtained from sorghum, maize, rice, cotton and sunflower, whereas soybean, wheat and beans generally produce small quantities.

Traditional varieties often yield greater quantities of residues than improved varieties, especially those of short stature and high harvest index. Most information on the optimum quantity of crop residues to be left on the soil surface is based on the amounts needed to reduce soil losses to acceptable levels on different slope gradients, rather than the amounts needed to maximize rainwater infiltration. Data exist showing that cover is less effective in reducing runoff than soil losses, but there is little information on the influence of cover on infiltration and runoff, especially on 20 to 50 percent slopes, which are commonly cultivated by small-scale farmers. Usually, a minimum value of 70 percent surface cover - equivalent to 4-6 t/ha of maize straw for example - should be adopted.

The quantity of residues remaining during the cropping season is also influenced by the rate of residue decomposition. Nitrogen-rich legume residues, such as those from beans and soybean, decompose much more rapidly than nitrogen-poor cereal straw and other residues with high

C/N ratios. On the other hand, legumes used as a cover crop can provide a weed-smothering cover, protection from raindrop impact, and important additions to organic matter. Harvesting procedures can drastically affect the quantity of residues remaining in the field.

The widely acclaimed success of conservation agriculture is mainly attributed to improved surface porosity that results in increased infiltration and reduced runoff, and a greater water availability to crops. As additional benefits, conservation agriculture also lessens evaporation losses, reduces erosion, enhances earthworm activity and soil structure, improves soil fertility and lowers labour, machinery and fuel costs. With time, yields increase substantially provided crop rotations are well designed and include leguminous crops or cover crops. When compared with only applying a soil cover (mulches, crop or cover crop residues) in a conventional system, no additional time is required for land preparation in CA (apart from herbicide application in some cases), which allows earlier sowing and all the advantages that this confers. Consequently returns to labour are substantially increased.

There is evidence that the yield of a crop is significantly higher when sown directly into the residues of a previous crop than when it is sown in a previously tilled soil to which the same quantity of crop residues are applied as a mulch. This is attributed to the benefits of little soil disturbance: the soil structure created by the root channels from the previous crops as well as by the biological activity of earthworms and other soil fauna facilitate deeper rooting and enhance the infiltration and percolation of rainwater.

Conservation agriculture principles are implemented optimizing the soil as a dynamic habitat for roots as follows:

— Residues of crops and of cover crops are distributed evenly and left on the soil surface.

— Once the soil has been initially brought into good porous condition, no implements are used to turn over the soil, to cultivate it or to incorporate crop residues.

— Weeds and cover crops are controlled by slashing with a knife roller or by preplanting application of a non-polluting desiccant herbicide.

— A specialized planter or drill cuts through the desiccated cover, slotting seed (and fertilizer) into the soil with minimum disturbance.

— Crop rotation is fundamental to zero tillage. It promotes adequate biomass levels for permanent residue cover and assists in control of weeds, pests and diseases. Rotations also ameliorate soil physical conditions, recycle nutrients and can fix atmospheric nitrogen. In semiarid conditions, appropriate crop rotations involving deep-rooting crops can also make still better use of residual soil moisture.

— As a result, soil erosion is reduced by about 90 percent and soil biological diversity maximized

In such systems soil damage is reduced and recuperation of soil architecture is much more quickly achieved than by unimproved fallow systems. Appropriate crop rotations are as important as the soil cover and no-tillage practices. Grasses, in particular, increase the aggregation and stability of soil particles which provide a range of small voids resulting in increased porosity.

Residue-based zero tillage is implemented gradually on structurally damaged soils. At the start, tillage with tined equipment (scarification) can be used to break up the underlying pan and let more rainwater back into the soil, while leaving some of the plant remains on the surface. In this way the soil is opened up and the previous crop's residues are incorporated. It may necessary to start renovating the soil by enabling more rainfall to become soil moisture, but too frequent scarification can also damage soil architecture because of the shattering effect on soil structural units.

Following the break-up of the underlying pan, strip cropping with a legume between rows of the main crop (e.g. maize) could be carried out. Finally a complete cover of crop residues without further soil disturbance by tillage could be established. The residues change overtime from being a protective cover to becoming an integral component of the soil. In the process, the worms and other soil mesofauna burrow within the soil seeking food and thereby provide channels and biopores through which air and water can move easily.

EFFECTS OF CONSERVATION AGRICULTURE

Effects on crop yields

Farmers' own experiences confirm what was anticipated by results of two six-year experiments with wheat and soybean between 1978 and 1984, comparing effects of conventional tillage, minimum tillage/scarification and zero tillage.

Effects on soil moisture

It might be expected that zero tillage would be no better than scarification in increasing moisture in the soil, but this is not the case. This shows changes in levels of soil moisture under wheat, at three depths, under conventional soil preparation, scarification (minimum tillage) and zero tillage, during the crop's vegetative stage in the 1981 growing season. Plant-available moisture was greater and water stress, due to drought, shorter under zero tillage than under the other methods.

Where the cover of residues was similar, the percentage of rainfall which infiltrated into scarified and zero-tilled soil differed by only 2-3 percent.

The fact that differences in the three-dimensional arrangement of the root habitat contribute to differences in root growth and function, even though soil moisture conditions may be almost the same, has profound implications. The best conditions for root growth and function appear to be where there has been no disturbance by tillage implements and where soil organisms are doing the work of burrowing, transforming and aggregating soil constituents. It may also be that differences in soil moisture inferred from runoff measurements under different tillage treatments may be insufficient to explain differences in root measurements and in final yield.

A pioneer farmer in Paraná, Brazil, whose soil conditions have been monitored from 1978 to the present, has kept detailed yield records. These show that under residue-based zero tillage, yields of both maize and soybean have been rising and have become less variable from year to year.

Effects on some other soil health indicators

The impacts of zero tillage (ZT) and conventional tillage (CT) on soil health are shown by comparing some soil indicators for both systems:

— diameter and stability of soil aggregates

— soil organic matter content at 20 cm depth

— number of earthworms

Saturnino and Landers measured the number of maize roots in each 10 cm layer of soil to 1 m depth after 15 years of constant treatment (zero tillage and conventional tillage). Zero tillage and crop rotation favour recycling of nutrients and better soil structure, resulting in better root development and higher production.

A research report from 1983 showed similar differences in root distribution of soybeans. While the total number of roots was the same to 1 m depth, they were more evenly distributed down the profile with zero tillage than with conventional tillage.

Effects on erosion and runoff

Conservation agriculture compared with conventional tillage results in markedly reduced soil erosion and runoff, as shown in results from Brazil and Paraguay. This effect is attributed to the increased soil porosity beneath residues due to biological activity. Note the saving of 441 mm water by reduction of runoff in southern Brazil and 186 mm in central Brazil.

In the Municipio of Tupanssi in Paraná it was reported that, following adoption of residue-based zero tillage, the turbidity of the river water has fallen from an index of 8 000 to 80 (author's field notes). A group of farm families, whose houses were on the slopes of cultivated fields recently transformed by zero tillage, said they were pleased that the runoff water and sediment no longer rushed down the hillsides into their houses, damaging the rugs and carpets on the floors. (author's field notes).

If runoff and erosion are symptoms of soil misuse, the major reduction in both occurrences signifies that their causes must have been significantly reduced.

Effects on catchment hydrology

An example of positive changes in catchment hydrology is provided by a representative catchment near Toledo in Paraná, Brazil. Soon after the adoption of zero tillage on rolling wheat lands, farm families observed that a pond which formerly had been dry for much of the year filled with water and hydrophytic vegetation took hold again. Further down the catchment, the river, which had ceased to flow in the dry season began to flow again throughout the year, so that a small farmer on its banks was able to improve his livelihood by investing in irrigation equipment and in excavating fishponds. This farmer now keeps fishponds full of water through the year and charges people to come fishing for fun at the weekends.

Effects of zero tillage systems on farm economics

Farmers have responded to the economic benefits of zero tillage. Yield increases of 20 percent or more, coupled with reduction of production costs by a similar percentage, have had positive effects on farm income. Savings of time and labour have contributed to improvements in farm families' livelihoods.

For instance in Paraguay, on farms using conventional tillage systems, severe losses of soil, nutrients and organic matter were seen as a root cause of declining yields of a range of crops. Some farms had adopted zero tillage, others not. Farm records over 10 years were used to construct economic models and indicators of differences. On representative mechanized 135 ha farms growing rotations including oats, soybean, sunflower, maize, wheat, crotalaria, vetch with zero tillage (ZT) farm incomes rose while those using conventional tillage (CT) for rotations with soybean, oats, wheat, maize fell. The returns on capital increased on farms using zero tillage, but declined on those using conventional tillage. Reduction of tractor-hours, reduced use of fuel and lower costs of repairs, etc. contributed to the economic benefits of zero tillage on these farms.

The small-farm study illustrates that zero tillage is not only financially attractive to small farmers but also has high economic pay-off for the nation. In Paraguay it has been estimated that for 1997 the national economic benefit due to the adoption of zero-tillage systems reached US$941 million. These included the saving in nutrients lost from soil from erosion, plus the costs saved in reduced tractor hours, less fuel and fertilizer.

The 1980 agricultural census of Paraná State, Brazil showed that there were over 6 million ha of annual crops. A 1989 report indicated the annual benefits if residue-based zero tillage systems were to be applied to the full 6 million ha.

Observations about Residue-based Zero Tillage Systems

From the full application of both the concepts and integrated techniques of residue-based zero tillage, farmers have achieved many direct and indirect benefits, often recorded together on individual farms.

On-farm benefits included:

— marked and rapid increase of organic matter content in upper layers of soil and increased biodiversity, number and activity (of earthworms, fungi, bacteria, etc.) in the soil;

— better soil structure and stability of soil aggregates; significantly higher infiltration rates; soil loss reduced by over 80 percent, runoff by 50 percent or more; more intensive but safe use of sloping areas made possible;

— increase in nutrients stored, greater availability of P, K, Ca, Mg in the root zone; less fertilizer needed for same result;

— better germination and development of plants, better root development and to much greater depth; better resilience of crops in rainless periods due to increased water holding capacity;

— yields often higher, typically + 20 percent for maize, + 37 percent for beans, + 27 percent for soybean, + 26 percent for onions; with less year-to-year yield variation;

— reduced variations of soil temperature during the day, with positive effects on plants' absorption of water and nutrients;

— less investment and reduced use of machinery and animals in crop production; reduced costs for labour, fuel and machinery-hours perceptible within 2 years. Operational net margins per ha rose by between + 58 percent and + 164 percent, because of combination of lower cost of production and increase in yields, which provides greater resilience against falling market prices and bad weather;

— greater flexibility in farm operations especially over optimum dates for planting; increasing possibilities for diversification into livestock, high-value and different crops, vertical integration into product processing and other activities; improved quality of life.

Off-farm benefits widely noted by rural agency staff and others, included:

— flooding risks reduced by 30-60 percent due to greater rainfall infiltration and delays to overland flows. Extending the time of concentration; better recharge of underground aquifers, improving groundwater reserves and dry season flow in springs and streams;

— less herbicide use after first years; less pesticide use, more recycling of animal wastes; reduction of pollution and eutrophication of surface waters by agricultural chemicals carried in surface runoff and eroded soil; less sedimentation and infrastructure damage, e.g. silting of waterways, large dams. A conservative estimate for the Cerrado region was given as US$33 million per year;

— reduced water treatment costs (ca. 50 percent) due to less sediment, less bacterial and chemical contamination;

- savings of up to 50 percent in costs of maintenance and erosion avoidance on rural roads;
- reductions in fuel consumption of 50-70 percent or more and proportional reduction in greenhouse gas emissions;
- reduced pressure on the agricultural frontier and reduced deforestation by high-yielding, sustainable conservation agriculture and increased pasture carrying capacity through rotation with annual crops;
- enhanced diversity and activity of soil biota;
- reduced carbon emissions through less fuel use and enhanced carbon sequestration by not destroying crop residues and increasing, rather than losing, soil organic matter.

The zero tillage systems of Latin America thus are not only a great improvement on former tillage-based systems, but also have major off-site and national benefits, to which improvements in soil moisture management make a large contribution. The effects are illustrated by the colour of the water going over the Iguassu Falls in southern Brazil (Plates 78 and 79). By chance these two Plates were taken from the same viewpoint 7 years apart, one in the wet season when high runoff also transported much eroded soil, the other in the dry season when water that had seeped down through the soil to the groundwater provided the dry-season flow.

Constraints of Conservation Agriculture

Conservation agriculture has been successfully employed in subhumid as well as humid climates, but there are still some constraints in semiarid environments that may hinder its immediate application. Typical of these constraints are:

- shortage of water limiting crop and residue production;

— insufficient residues produced by the economically or socially important crops and lack of knowledge of suitable cover crops;
— sale or preferential use of crop residues for fodder, fuel and building materials;
— inability to control livestock grazing, especially in areas where communal grazing is traditional (tenant farmers are often obliged to allow the landowner's cattle to graze the residues after harvest);
— inability to control residue consumption by termites;
— insufficient money or credit to purchase appropriate equipment and supplies;
— lack of knowledge of conservation agriculture by extension and research staff.

A number of approaches have been explored and are being tested to overcome these constraints. In situations where crop residues are preferentially used as fodder, additional new sources of fodder may be produced, provided they can be protected from grazing by, for example, live fences. Hay or silage may be produced as additional dry-season fodder from improved pasture species, or from forage trees or crops of high biomass grown specifically for this purpose. Forage trees can be established as live fences along farm and field boundaries, and forage grasses may be produced as live barriers, on bunds, and along field boundaries and roadways. In Bahir Dar, Ethiopia, farmers are increasing fodder production by undersowing forage legumes in other crops, establishing forage strips between arable crops, and by oversowing mixtures of legume seeds on grazing areas.

Certain crop sequences are less suited to direct sowing into crop residues because of the likelihood that weed, pest or disease problems will become intensified by being transmitted from one crop to the next. Examples of less suitable crop sequences and their specific problems encountered in eastern Bolivia are:

— wheat every year - disease problems;

— soybean every year - pest and disease problems;

— soybean-sunflower sequences - disease problems;

— maize-sorghum or sorghum-black oats - weed and pest problems;

— sunflower-cotton - the problem of volunteer sunflower weeds;

— bean-soybean sequences - pest and disease problems.

Weed problems may also be caused by volunteer germination of the previous crop; for example, sunflower volunteers can be particularly difficult to eradicate. To avoid such problems, appropriate crop rotations, acceptable to the farmers, must be selected.

In environments where there are many constraints to the introduction of conservation agriculture, a pragmatic, phased approach may be the most feasible, in which individual constraints are progressively overcome until an appropriate system of conservation agriculture can be fully implemented. This may require the planned introduction of measures such as improved grass species and fodder trees, hay and silage production, live fences, stall-fed livestock, improved crop rotations with cover crops, formation of farmers' associations, credit supply and local or international training visits for farmers, extension and research staff.

The introduction of conservation agriculture is unlikely to be immediately successful on seriously degraded soils with surface crusts, compacted layers, low fertility or severe weed infestations unless these problems are first overcome by appropriate remedial actions. Hardsetting soils may not be immediately suitable for conservation agriculture because of the difficulties of overcoming soil compaction problems and maintaining good soil porosity within the topsoil and subsoil. Consequently crop rooting is frequently restricted to shallow depths. In this case, deep tillage followed by the establishment of cover crops prior to introducing conservation agriculture, and then the adoption of crop rotations that produce large quantities of residues, will progressively improve the physical condition of these soils and make conservation agriculture possible.

Conservation agriculture is less likely to be successful in poorly drained soils because the added residues will intensify anaerobic conditions, in which toxic substances harmful to crop growth may be produced.

The cost of no-till planters and seed drills needed for direct sowing may be a major constraint for mechanized farmers, unless it is possible to modify their existing seed drills and planters. For small farmers, hand tools and animal-drawn equipment exist and local blacksmiths can often adapt them, provided they have access to information and samples.

2

Conservation Agriculture and Sustainable Crop Intensification

In recent years, the spread of conservation agriculture (CA) has revealed to be a sustainable way to intensify crop production and sustain rural livelihoods in several African countries. Indeed, the potential benefits associated with the use of conservation farming practices are many. Long-term yield increase and output stability can be achieved while at the same time stopping and reversing land degradation. Larger outputs are often obtained by employing relatively fewer inputs, thereby reducing costs. Compared to conventional tillage methods, CA thus leads to higher net profitability, greater environmental sustainability and—especially important in Africa—higher food security. Furthermore, conservation farming techniques which rationalize the use of labour are particularly helpful in those rural areas where migration and health emergencies have reduced the labour supply and contributed to the increasing "feminization" of the agricultural.

The present chapter reviews and analyses the information collected under a baseline survey, undertaken in Lesotho with the aim to assess the potential costs and benefits associated with the adoption of a planting basins system, locally called likoti.

Lesotho: A Context of Growing Vulnerability

Lesotho is a small, landlocked country of about two million people, of whom 76% are rural. With a GDP per capita estimated at US$1,541 in 2007 and 68% of the population living below the national poverty line, it is one of the world's poorest countries. Its economy is based on limited agricultural and pastoral production, light manufacturing (led by export-oriented garment factories owned by East Asian investors) and remittances from migrant labour (albeit declining compared to the past).

Even though social indicators are generally better than the Sub-Saharan Africa (SSA) average, in 2009 the United Nations Development Programme (UNDP) ranked Lesotho as 156th out of 182 countries based on its Human Development Index, and as 106th out of 135 countries based on its Human Poverty Index. The delivery of social services is extremely weak: health personnel are in short supply, health centres are not adequately equipped, and schools lack teaching materials. Over the last ten years, a major health problem has been the increasing spread of HIV/AIDS. According to UNAIDS, in 2008 23.2% of the population aged 15-49 was infected, one of the highest figures in SSA.

The spread of HIV/AIDS, along with high unemployment rates, mainly due to the retrenchment of many Basotho miners, are among the most important causes of poverty and vulnerability. Along with the migration towards urban and peri-urban areas, and the absorption of many young female workers by the textile industries, these trends are affecting the traditional social structures within the household and at village level. As a result, the social protection mechanisms which so far have helped the Basotho people cope with shocks and stresses are in decline. At the same time, public welfare policies have failed to take over these tasks.

The economic and social transformations described above occur in a risk prone environment, where the scarcity of natural resources, especially fertile land, is at the same time a cause and a consequence of poverty. Lesotho's ecology is fragile because of its mountainous topography (it lies on a high plateau that rises from 1,500 metres in the west to 3,350 metres in the east), the thin soil layer and the limited vegetative cover. Under such unfavourable conditions, the high pressure of human and livestock activities on the land has led to major environmental problems. Forests, as well as pastures, are progressively disappearing. At the same time, the impressive extent of soil erosion increases river siltation and gully erosion.

Naturally acid soils with poor contents of phosphorus and organic matter, along with land degradation and extreme climate conditions, have steadily reduced the potential agricultural output. The sector's share of GDP has fallen from 50% to about 15% since the mid-seventies, while yields have fallen by about two-thirds during the same period. The agricultural potential is limited not only by the scarce availability of fertile land and by external shocks, but also by the social and economic transformations which are limiting the access to physical as well as social assets.

As highlighted by Boehm, farming in Lesotho is "an activity characterised by a high level of sociality". Since very few farmers own all the necessary assets and means of production, Basotho depend on various forms of co-operation and sharecropping agreements (seahlolo or lihalefote). In order to successfully achieve these agreements, farmers need to use a number of "social skills", including trust, reliability and reciprocity. In other words, they need to rely on social capital. But, as has been already mentioned, social capital in Lesotho has been increasingly affected by unemployment, the associated increase in income poverty, and the spread of HIV/AIDS. All these factors limit the effectiveness of social assets and sharing mechanisms, thereby affecting the capability to sharecrop and ultimately to farm. Indeed, an assessment of the food security situation undertaken by the Forum for Food Security in Southern Africa, found that the recent food crises stemmed only partially from crop failures and adverse weather conditions. Rather they reflect long-term, latent food insecurity, in turn caused by poverty (lack of physical assets), deteriorating social capital, and negative social and economic trends due to migrations, retrenchments, and the HIV/AIDS pandemic.

All the problems mentioned so far are exacerbated by poor governance and inefficient governing institutions. Even though corruption remains low compared to other African countries, poor law enforcement, insecure property rights, inadequate delivery of public services, and inadequate local government (including problematic integration of traditional and modern institutions), slow down economic growth and development, and discourage people's participation in civic and political life. As a result, Basotho live in a context of growing vulnerability, reflected in increasing poverty and inequality, deteriorating health conditions (including low standards of food and nutrition security), and increasing exposure to external shocks and changing climatic conditions.

Diffusion of Conservation Agriculture as an Innovative Process

In Lesotho, agriculture consists primarily of maize and (to a much lesser extent) wheat mono-cropping. In spite of abundant and irregular rains, rainwater harvesting methods are rarely practised. Agricultural productivity is highly variable (especially due to erratic precipitations), and it has steadily declined over the latest 30 years—maize yields have fallen from an average 1,200 Kg per hectare in the mid

1970s to a current 450-500 Kg per hectare in most of the districts. Nonetheless, agriculture remains a source of livelihood for the vast majority of the population, most of which is engaged in subsistence farming.

In spite of recent attempts to strengthen the invovement of the private sector and encourage diversification into high-value export products (such as the Agricultural Sector Adjustment Programme assisted by the African Development Bank in 2000 progress has been limited by the poor prospects of profit as well as by ineffective agricultural development policies. The livestock sector provides a significant proportion of rural income (usually for better-off households) and is well integrated in the national and the regional economy through the export of wool and mohair. However, the importance of livestock has also started to decline due to the recurrent droughts, poor animal quality and inadequate disease control.

The potential opportunities for the agricultural sector to develop and contribute to the country's economic growth are contested. This is in part due to contradictory figures on agricultural production (in the latest seasons, estimates of the cereal gap varied by over 300% depending on the source). Whatever the precise data, it is hardly contestable that agricultural production is in sharp decline. Shortage of arable land, mainly due to land degradation, and erratic climate are commonly mentioned as the most determinant factors. However, these problems have affected the country since it was a British protectorate in the nineteenth century. Therefore, it is most likely that the main causes of this decline depend on the farmers' limited or inadequate capability to deal with and adapt to the environmental conditions.

The abandonment of the fields by migrant workers, and the consequent scarce investments, have limited the adoption of products and technologies suitable to the local conditions, and ultimately hampered the growth of productivity. At the same time, the use of conventional tillage methods and of intensive agricultural practices, often promoted by development assistance programmes, have contributed to increased land degradation and lower soil fertility. More recently, the poor performance of agriculture has been linked also to the high rates of poverty and vulnerability (discussed above) which affect the economic as well as the social capabilities needed to farm. Finally, in some cases, poor crop production has been the paradoxical outcome of policies aimed at encouraging food production—such as subsidies and emergency interventions—which resulted in late plantings or disincentives to plant.

Of course, none of these factors, alone, is the unique or main cause of the dramatic decline of yields and output. A complex combination of interrelated factors has contributed to the current situation. With regard to the future, some see the agricultural sector as a disaster, while others recognise the potential for increasing agricultural productivity and stress the role of agriculture in combating poverty and enhancing food security. Those who support the potential role of agriculture in development recognise the need to face—and overcome—several challenges. Among the most important of these is the need to foster agricultural development through the promotion of more sustainable, ecologically friendly practices, such as soil and water conservation, reclamation of limited areas of degraded land for intensive food production, and mixed and low external input farming.

Even if, at least in the medium term, it seems unlikely that agriculture will be the driver of economic growth and provide significant numbers of jobs, proper policy options and interventions aiming at enhancing the availability of food could stimulate local agricultural markets, and contribute to creating employment and increasing wage rates through higher productivity. In order to boost agricultural yields and stabilize outputs, however, appropriate solutions should especially focus on environmental as well as social sustainability. In fact, in such a risk prone natural environment, conventional tillage methods impose a severe stress to the soil and decrease crop productivity. Furthermore, practices which rely on expensive purchased inputs and mechanical implements increase farmers' vulnerability to external shocks. On the other hand, conservation agriculture (CA) can provide an effective solution to reversing the spiral of declining productivity caused by land degradation and extreme environmental conditions. In addition, some CA practices are particularly suitable to small-scale and poor resource farmers..

Since 2000, a growing number of local and international actors—including, among the others, the Food and Agricultural Organization (FAO), the World Food Programme (WFP), the National University of Lesotho (NUL) and several NGOs—have started to promote conservation agriculture in Lesotho. The adoption of conservation farming as a strategic means to increase and stabilise agricultural production as well as to prevent and reverse soil erosion has been explicitly mentioned in the Lesotho Food Security Policy and Strategic Guidelines (Ministry of Agriculture and Food Security (MAFS), and in 2006 the FAO Representation in Lesotho launched a Conservation Farming Network Group (CFNG) to encourage the adoption of water and soil conservation technologies, facilitate the exchange of knowledge among different actors and provide them with updated information about the CA practices that were promoted in the country and the associated outcomes.

According to the information collected through the CFNG, several different CA practices have been implemented in Lesotho in the latest years. Between 2005 and 2006, the MAFS set up field trials to support the adoption of animal and tractor-drawn no-till planters under "block farming" agreements, while FAO organized training and demonstration sessions to teach farmers to use simple no-till equipment such as jab-planters. Furthermore, some NGOs were committed to (re)introduce sustainable agricultural practices such as the Machobane Farming System. However, all these efforts have achieved mixed results. The conservation farming practice that so far has shown the highest potential is a planting basin system, locally called likoti. The practice was introduced in Lesotho mainly due to the commitment of two Christian organisations: a local association called Growing Nations, supported by Africa Inland Mission (AIM), and Rehoboth Christian Church. Up to today, likoti has been by far the conservation technique most adopted countrywide and the reason of this success largely lies on the inspiration and motivation of its promoters.

The Practice of Likoti

Rev. August Basson is an AIM missionary who moved with his family to the harsh mountainous area of Tebellong, in Qacha's Nek district, in 1993. For many years, he was trying to improve local agriculture by investing donors' money in tractors, greenhouses, and inputs, but the yields never repaid his investments. Furthermore, most Basotho people could not have afforded such equipment. The Pastor then switched to testing farming practices that relied on low external inputs, which were more suitable to the local socio-economic conditions. He also realized that the tillage methods that were in use in Lesotho were exacerbating soil erosion and land degradation. Eventually, in 2000, Rev. Basson went to South Africa to learn more about conservation agriculture.

Back to Lesotho, he developed a planting basins system adapted to the local conditions and started to promote it with a Sesotho name, likoti, which means "holes". According to the likoti method, pits are about 15x30cm large in diameter and 15-20cm deep (or smaller), and they are dug in a 75x75cm grid. A small quantity of fertilizer (either inorganic or organic) and seeds (the number depends on the desired crop density) are placed in each basin and covered with soil. Additionally, farmers should leave crop residues on the field as mulch and practise crop rotation and inter-cropping. The next season farmers can plant in the same pits without turning the soil up. The Likoti method was originally deployed in the production of maize and beans. However, innovative farmers have used it also to produce other crops such as sunflowers, sorghum, potato and tomato.

In order to diffuse the conservation farming technique, Rev. Basson founded a Lesotho based charity, Growing Nations. He hired fields to set up demonstration plots and organized several training sessions. It took a couple of years for the first group of likoti farmers to become confident with the new practice. In the meanwhile, Pastor John Mokoena and Rev. Pete West, from Rehoboth Christian Church, also started to promote CA in the northern district of Botha-Bothe. In 2001, Brian Oldrieve, a pioneer of the planting

basin system in Zimbabwe and other african countries since the Eighties, came to Lesotho to provide training both in Tebellong and Makhoakhoeng. Since 2002, the practice has captured the interest of more NGOs and international organizations, such as German Red Cross, the Food and Agriculture Organization (FAO) and the WFP, that provide different kinds of incentives to farmers who adopt it.

According to the information provided by CFNG members, in 2006 about 500 households were practising likoti in the southern districts of Qacha's Nek, Quthing and Mohale's Hoek, whereas the number of CA farmers in Botha-Bothe and Berea districts, in the northern lowlands, was about 350. Since then, the number of likoti farmers has steadily increased. Even though precise figures are not available, WFP alone estimates that so far about 5,000 households—or 1.5% of the rural households—have adopted likoti with its support in different districts. Considering that the farmers supported by WFP cultivate on average 1.63 ha of land, currently there are about 8,163 ha of land under conservation agriculture (2.5% of the total arable land). However, these figures do not include the farmers who have adopted the likoti practice with the support of other organizations such as Growing Nations and FAO as well as those farmers who have adopted the practice on their own accord.

Source of Data, Field Survey and Methodology

This chapter analyses and discusses the information collected in 2006 under an FAO initiative entitled "Monitoring and evaluating the impacts of Conservation Agriculture (CA) in Lesotho: implementation of a comprehensive baseline study comparing conservation and conventional practices at small-scale farming level". The baseline survey was implemented under the FAO project OSRO/LES/503/UK—Support to vulnerable rural households in Lesotho—through a working arrangement between FAO Representation in Lesotho, the Department of Economics of the University of Roma Tre (Rome, Italy), the Department of Soil Science & Resource Conservation of the National University of Lesotho (NUL, Roma, Lesotho), and the Faculty of Lifescience of Copenhagen University (Denmark). The specific objective of the survey was to assess the socio-economic impacts of likoti on small-scale farmers, with a special focus on agricultural outputs, food security status, and economic and environmental sustainability.

The baseline study was implemented from January to October 2006. Two sub-sample populations of CA and conventional farmers were monitored through a household survey. The areas included in the survey were: Makhoakhoeng and Ha Mamathe in the western lowlands of Botha-Bothe and Berea districts, respectively, and Tebellong and Tsoelike in Qacha's Nek, in the mountains. The sites were selected from those where CA had been practised for a longer time (at least two agricultural seasons). In total, a sample of 229 farmers (117 CA and 112 conventional) was interviewed in two phases—before and after the harvest. The sub-samples represent a cross section of households in the selected sites. CA farmers were selected randomly amongst those who participated in training or other CA related initiatives (the complete lists were provided by local organizations promoting CA). Conventional farmers were selected partly randomly and partly purposively, in order to compare soil structure and crop yields in 'conservation' and 'conventional' fields.

The questionnaire submitted during the first phase was made of five sections: background household information (household composition and wealth status); food availability and food security; farmers' participation in associations and community organization; agricultural production activities; knowledge and perception of CA practices. During this phase, 165 composite soil samples were collected and a random selection of 123 was sent to the Department of Agriculture in Cedara, KwaZulu-Natal, and analysed for soil fertility and soil texture. The questionnaire submitted during the second phase included information on: agricultural yields and profitability; food security status; household vulnerability and community cohesion. During this phase, the yields obtained by a smaller sub-sample of farmers were recorded.

The information collected through the questionnaires allowed two datasets to be organised, comprising of about 300 variables on demographic and social features, household composition, wealth, assets, food security, community organization, social capital, agricultural production activities, knowledge and perception of CA and agricultural yields. Part of these data were analysed and the results were discussed in a document prepared for FAO Representation Lesotho and funded by DFID. Subsequently, the author undertook additional analysis under her Ph.D. research with the aim to evaluate whether and how social capital affected the adoption of CA as an innovative agricultural practice in Lesotho.

Characteristics of the Surveyed Sites

Lesotho is divided into four agro-ecological zones - Lowlands, Foothills, Mountains, and the Senqu River Valley. The Lowlands cover the western part of the country and occupy about 5,200 km, or 17% of the total surface area. In spite of its narrow extension, this region supports more than half of the national population, constitutes 70% of the limited arable land, and provides most of the available non-agricultural employment. The Northern lowlands are the most agriculturally productive and receive more reliable rainfall. The Southern Lowlands are generally warmer and dryer. By contrast, the mountainous eastern side of the country is sparsely settled, arable land is scarce and communities are much more isolated from urban services and markets.

Qacha's Nek district is located in the south-eastern highlands. Its land area is 2,349 km of which only 2,000 ha (less than 1% of total land area) are used for crop production. It has approximately 3.7% of the national population with a density of 34 people per km. The chieftainship of Tebellong lies at an elevation of 1,700 to

1,900 m a.s.l. while Tsoelike lies between 1,800 and 2,000 m a.s.l. Both areas are situated on top of the Senqu River Valley escarpment. The average annual rainfall is 800 mm. Summer temperatures range from 18°C to 25°C while the winters are cool (0° to 15°C) with frequent snowfall in July and early September. Crops planted include maize, beans and sorghum. Wheat is mainly grown in the highlands of the district.

Most of the surveyed sites in Botha-Bothe and Berea districts are located in the Northern Lowlands. The land area of Botha-Bothe is 1,767 km of which 7,763 ha (less than 5% of the total land area) is used for crop production. The population accounts for 5.9% of the total Lesotho population with a population density of approximately 72 people per km. Qholaqhoe is the principal village in the Makhoakhoeng area. It lies on the borderline between the lowlands and the foothills of the northern part of the district at elevations between 1,500 and 2,000 m a.s.l. and occupies the upper slope positions of the sandstone escarpment above the Caledon river flood plains. The average annual rainfall is 800 mm with 60% falling during the summer months of December and January. Summer temperatures range from 25°C to 34°C while the winters are cold (-2 to 14°C) with occasional snowfall around June and July. The major crops include cereals—maize and sorghum—beans and potatoes. However, winter precipitation in good years allows for a winter cropping season with wheat and peas as the main crops.

The land area of Berea district is 2,222 km of which 27,158 ha (about 8% of the total land area) is used for crop production. The population accounts for 13.9% of the total Lesotho population and the population density is of 135 people per km. The village of Ha Mamathe is located in the northern lowlands of the district at an elevation ranging from 1,500 to 1,800 m a.s.l. The area is characterized by the same agro-ecological features and climatic conditions present in Botha-Bothe. Maize is by far the most grown crop. Other crops include beans and sorghum.

The consequences of soil erosion and land degradation are clearly visible both in the mountains and the lowlands: overgrazing of range lands, uncontrolled burning and cultivation of steep slopes without the implementation of proper conservation measures have progressively denuded the landscape. In all the surveyed sites, farmers reported to observe a steady decline in their fields' fertility and attributed it to the fragility of the local environment. Indeed, as was mentioned earlier, in Lesotho land degradation (due mainly to the overexploitation of the natural resource base) and chemical soil depletion (loss of organic matter, nutrients and acidification) exacerbate each other in a vicious circle which increasingly exposes the landscape to water and wind erosion.

CA at the Surveyed Sites

The practice of likoti was first introduced in Tebellong in 2000 by Pastor August Basson and his locally based association, Growing Nations. In 2003 FAO started to support Growing Nations in providing seeds and fertilizer to farmers who adopted the practice. In 2005, WFP also started to support the diffusion of conservation farming, by providing Food for Work (FFW) to farmers who attended training and worked in the fields of the 11 local trainers who in the meanwhile had taken over Rev. Basson's duties. A Masotho Pastor, Rev. Rantimo, has been the pioneer of likoti in Tsoelike. Just as Rev. Basson has done, Pastor Rantimo started to train farmers elsewhere in the country (in collaboration with Dorcas Aid, a Dutch NGO), while a local trainer, Mr. Isaac Sehahle, took over his duties. Pastor John Mokoena and Rev. Pete West, from Rehoboth Christian Church, started likoti in Makhoakhoeng in the cropping season 2001/02. They immediately decided to supply the farmers who joined their training programme with seeds and fertilizers, because most of them lacked production assets and money to buy inputs. In 2006, there were 14 "CA leaders" in Makhoakoheng and Ha Mamathe, each being responsible for a village or an area.

During the survey, two workshops were organized with the local trainers—one in Tebellong and one in Thaba Kholo (Makhoakhoeng)—with the aim of discussing the opportunities and the challenges associated with the diffusion of likoti, as well as to better understand their training methodology. In Tebellong and Tsoelike CA trainers use to follow up the training sessions by visiting the farmers on average three times per month and organizing open gatherings if they feel that there is any issue or problem to be discussed. By visiting farmers while they are at work in their fields, and organizing the meetings there, they often capture the interest of the other farmers who are nearby. Apart from teaching the principles of likoti, trainers encourage household members to work together, as well as to organize work parties with other farmers. CA leaders in Makhoakoheng and Ha Mamathe go to visit the trainees on average once a month and do not organize regular meetings, although they describe individual interaction as quite constant. Even though they encourage CA farmers to work collectively, work parties are very rare. According to their view, farmers are not cooperative and they do not trust others' commitment.

Characteristics of the Sample

In order to assess the distribution of welfare among the respondents, two indexes were formulated based on the existing literature on livelihood strategies and well-being in Lesotho: an Asset Index and a Capabilities Index. The former measures the endowment with productive assets, which includes livestock, land and other productive means. The index can be considered a proxy of the wealth of the households, even though it does not include other economic resources and monetary earnings. The Capabilities Index is built by synthesizing variables such as the availability of a salary and other formal income sources, the ownership of a tractor, the capability to hire workers through matsema (collective wok), the presence of disabled members in the household and the household dependency ratio (calculated as the ratio of household members more than 60 years old and less than 18 to the 19-59 year old members). It measures the household capability to generate welfare. Both indices range between zero and 100, and the respective distributions have been divided into quintiles in order to rank the households as very poor, poor, moderate, better off and rich.

On the basis of the Assets Index, most of the farmers can be classified as poor. However, the percentage of CA farmers classified as very poor (the lowest quintile) and poor (the fourth quintile) is higher compared to the conventional farmers. Consistently with these observations, the difference between the average Assets Index of the 'conventional' group (40) and the average Assets Index of the CA group (34.6) is statistically significant at 5% level ($t=2.4$; $t_{0.95}=1.65$). It is worthy to note that, within each sub-sample (or farmer category), female headed households (FHH) are less endowed with productive assets (i.e. a higher share is classified as poor) compared to male headed households (MHH). The wide majority of farmers can be classified as poor also on the basis of the Capabilities Index. About 10% of the FHH and 5% of the MHH in each sub-sample belong to the lowest quintile, meaning that they can not generate enough income to achieve minimum living standards. Taking into account the average Capabilities Index for the two farmer groups (41.5 and 39, for the conventional and the CA farmers, respectively), the differences between conventional and CA farmers are still present, although less significant ($t=1.35$; $t_{0.9}=1.28$).

These results are consistent with the figures available at national level, according to which the wide majority of Basotho families in rural areas are poor. In relative terms, some households can be considered better off because they rely on some kind of formal income or because they are more endowed with land and livestock. Others, and especially most vulnerable categories (elderly or sick people, orphans, widowed women), rely instead on a reduced or unstable livelihood basis, experiencing a declining welfare and food security status.

In the surveyed sample, most households can be classified as poor, but those practising likoti, and especially the female headed households, are even poorer. The presence of vulnerable categories amongst CA farmers is not surprising, since most of the projects promoting CA in Lesotho targeted disadvantaged households. The baseline survey was conducted over a period of six months during the same agricultural season, so it was not possible to include a dynamic assessment of the impact of CA on farmers' well-being status. However, the fact that most farmers continued to practise likoti also after the incentives had ended means that the technology proved to be suitable to their socio-economic conditions.

While physical assets seem not to be a determinant factor for the adoption of CA, literacy and education play an important role. The frequency analysis shows that the share of literate heads of the household plus those who completed the primary school is larger among the CA farmers. It is interesting to notice also that, while the women in the conventional sub-sample have the lowest degrees of education at all levels, the women in the CA sub-sample reported the highest percentage of literacy and accomplishment of primary school among all respondents.

These results confirm a recurrent finding in the literature, i.e. the critical relevance of education to innovation adoption and diffusion. Furthermore, the fact that the women of the CA sample are significantly more educated than the women of the conventional sample suggests that human capital represents a determinant asset, particularly when access to other assets is limited by socio-economic factors or gender bias, as still occurs in Lesotho.

The Impact of Conservation Agriculture on Sustainable Crop Intensification

As expected, the interviewees planted mainly maize, sorghum and beans. Most farmers in both sub-samples cultivate one or two fields (including both owned and sharecropped fields), the total land cultivated by CA farmers (0.77 ha) being on average smaller than the land cultivated by the conventional farmers (0.95 ha). So the size of the fields is usually very small and female headed households—both CA and conventional—plant even smaller fields.

Data on the outputs for each crop were collected through the second questionnaire and the yields were calculated in tons of maize/ha using appropriate conversion factors for beans and sorghum. Unfortunately, the quality of the information collected through the questionnaire was extremely low. A part from mistakes and missing answers that could occur due to the different methods farmers used to measure the output, respondents tended to declare smaller yields compared to those they actually got. In order to get a better estimation of the yields, the output was directly measured for a sub-sample of farmers. As a whole, data were recorded from 29 fields in Makhoakhoeng (14 likoti and 15 ploughed), 26 being comparable fields (fields located next to each other with similar soil characteristics). In Tebellong and Tsoelike (Qacha's Nek), yield records were collected from 24 CA farmers, whereas it had not been possible to directly measure the yields of the conventional farmers due to logistical and time problems. The visits to the farmers confirmed that, when asked, they often reported a lower output compared to what was then recorded in the store room. This is not surprising taken into account that in the Basotho culture people do not like to talk about their economic resources.

According to the records collected for this sub-sample of farmers, in Makhoakhoeng the maize yields obtained by using likoti (1.36 t/ha) were higher than those obtained in the ploughed fields (0.87 t/ha), the difference being statistically significant at the 0.1 level. In Qacha's Nek, CA farmers got about 0.73 t/ha of maize, which is more than three times the district average yield for that growing season - 0.2 t/ha, according to WFP and FAO.

The most important difference between the conventional and the conservation farming practices is the tillage method. All the households included in the CA sub-sample planted at least one field using likoti. Among the conventional farmers, the wide majority ploughed with animals. Indeed, hiring a tractor to plough may be relatively expensive: according to the prices recorded under the survey, on average a tractor owner charges M494 per hectare. Ploughing with animals does not imply any monetary expense but the household needs either to have its own cattle or needs to rely on some assets (land, money to buy inputs, mechanical equipment, and so on) in order to sharecrop with somebody who owns cattle.

Digging planting basins with a hoe is thus cheaper than ploughing, since virtually every household with one or more members capable to farm some land can afford it. On the other hand, the workload required by this planting system is much heavier. CA farmers reported that digging the holes and planting with likoti took on average 320 man hours per hectare, compared to 145 man hours/ha usually needed for ploughing and planting with animals. These figures are consistent with the information provided by the CA trainers, according to whom likoti farmers need on average 375 man hours/ha the first season and 275 man hours/ha or less from the second season onwards.

Furthermore, likoti farmers are taught to weed their fields three times per season. In fact, 63% of the CA respondents said to weed at least twice compared to 34% of the conventional farmers. Considering that the large majority of the farmers (90%) in both categories weeded by hand (which takes on average 150 man hours/ha), the amount of labour required to weed the likoti fields is considerably higher. However, as will be discussed later, labour requirements diminish progressively over time.

These findings suggest that, in order to fully understand the costs and the benefits associated with the CA practices, and compare them with those associated to conventional practices, it is necessary to consider not only the yields, but also other aspects, such as the economic profitability, the returns to labour and the social sustainability of the technique.

Economic Analysis And Labour Productivity

Both conventional and CA farmers used on average 17 kg/ha of seeds of maize, in accordance with the quantities recommended by the MAFS (15-20 kg/ha). Regarding the use of inorganic fertilizers, CA trainers recommend farmers to use eight grams per hole, which gives 140 kg/ ha, but farmers who reported to employ inorganic fertilizer used on average only 87 kg/ha. Conventional farmers used on average 105 kg/ha of inorganic fertilizer, which is also very low compared to the quantities recommended by the MAFS (300 kg/ha of 3:2:1 (25) for maize, sorghum and wheat). The cost of pesticides and herbicides has not been considered in the analysis because only a negligible share of farmers in both categories reported to be able to afford them. The fee to hire the tractor is included among the costs of the farmers who ploughed with the tractor and sowed using draught power

Since in Lesotho agricultural labour is usually provided by household members, it was not possible to consider wages as a direct cost. The opportunity cost of labour has been approximated to zero for two reasons: (i) almost all farmers are practising subsistence agriculture; (ii) there is a high rate of unemployment and lack of alternative off-farm jobs, especially in rural areas. As a result, no costs were assigned to labour intensive activities such as ploughing and sowing with animals, digging the basins and weeding by hand. On the other hand, given the high importance that labour requirements play in the adoption of CA practices, labour productivity has been analysed separately.

In all the surveyed sites, the yield obtained using likoti would give the farmers a significantly higher profit: in Quacha's Nek using likoti gives a profit of M1,065 per hectare (compared to a deficit from ploughing), while in Botha-Bothe it gives a profit at M1,420 per hectare, compared to M206 when farmers ploughed with the tractor and M700 when they ploughed and planted with animals. The analysis shows that in Qacha's Nek, farmers who ploughed incurred a loss. Indeed, the average yield of 200 kg/ha obtained in the 2005/06 season is very low, even lower than usual.

It could be asked why these households continue to farm using conventional tillage methods, even if they make a loss. First it should be recalled that in this analysis the estimation of the profit serves exclusively as a comparable measure of the benefits associated to the employment of different farming practices. To this aim, a monetary value has been attached to each item, even if in reality farmers neither disbursed nor received any money. For instance, some farmers (especially those who adopted likoti) received donations of seeds and fertilizers. On the other hand, many conventional farmers used their own seeds and manure thus had lower costs. Finally, concerning the value of the production, only the price of the maize grain was considered. But maize stubble also has a value as fuel and fodder, which is difficult to quantify. The existence of costs and benefits which can not be easily monetized—including traditional and cultural factors—may explain why farmers continue to cultivate even if they get such low yields.

So far the opportunity cost of labour has been assumed to be nil and wages have not been included as direct costs. However, especially considering that the workload required by the CA practices in the first seasons was reported to be one of the most important deterrent to their adoption, a thorough economic analysis has to consider also the productivity of labour.

In Qacha's Nek, where average yields are usually very low, conventional farmers made a loss, and therefore it is not possible to compare the labour productivity on the basis of the accounting profit. Comparing the different farming practices in relation to the output/labour ratio, likoti get the highest remuneration (2.36 M/man hour). In Botha-Bothe, the profitability of likoti calculated through the profit/labour ratio is almost the same as the profitability obtained by ploughing with animals and the double the profitability obtained by using the tractor. Therefore, the amount of labour required by likoti, although much higher compared to the conventional farming practices, is adequately compensated by the yield and return.

Considering that the figures used for the calculations refer to the workload necessary in the early adoption phases, which is expected to decrease over time, the profitability of CA is likely to augment in the subsequent seasons. On the other hand, if the labour available is not enough to accomplish all the farming activities on time, the output can be badly affected. It is therefore important that the household assesses carefully the labour force available in relation to the number and the size of the fields, in order to get an adequate labour remuneration. At the same time, improved farm management—to be achieved through an incremental shift to CA—helps the household to overcome the resource constraints. For instance, spreading the workload all over the dry season allows farmers to better manage the labour force available. Also frequent and timely weeding significantly reduces weed infestations over time.

Impact on Food Security

The social sustainability of likoti has been assessed also by monitoring the food security status of the surveyed households before and after the harvest. In both phases of the survey, the food security status was evaluated in relation to access and availability of food, frequency of meals and diversity of the diet. A food consumption score (FCS) was calculated to measure the diversity of consumption. Based on the FCS, the quality of the diet can be classified as low (FCS<10), adequate (10=<FCS<22) or good (22=<FCS<48). In both sub-samples, the majority of the households consume an adequate or a good diet; however, the average FCS obtained by those who fall in the adequate category is relatively small (average FCS=16) in relation to the top value of the range (FCS=22). In the case of Lesotho, a FCS equal to 16 means a basic diet of papa (maize meal) and moroho (leafy vegetables) and a few alternatives.

In the second phase of the household survey (just after the harvest), CA farmers reported a more diversified diet (i.e., they obtained a higher average FCS), while the conventional farmers' average FCS has slightly diminished. The differences are larger if the FCS is calculated excluding the consumption of fruit, which is widely available in rural areas in the summer, but is not in the winter after harvest. Without considering the consumption of fruit, both farmer categories improve the diversity of the diet, but the increase is much more significant for the CA farmers (19%) than the conventional ones (6%). This result would suggest that the availability of own production is more important for the CA sample and also that:

— CA farmers rely more on their own production, even though on average they cultivate smaller fields and employ fewer resources (the CA sample also give a higher importance to Own Production as main source of food).

— Considering that initially most of the CA farmers were involved in a CA project just because they had been targeted as vulnerable and food insecure (which was confirmed by the socio-economic analysis), the improvement in the FCS after the harvest would indicate that currently they manage to achieve an adequate food security status through their own production.

Impact on Soil and Water Conservation

The agro-environmental benefits associated with the use of CA practices are widely documented in the literature. The most important ones can be summarized as follow:

— Reduced tillage improves soil structure and stability and leads to the progressive suppression of weed growth as the seed bank of weeds in the soil decays.

— Crop residues left on the soil surface protect the soil from wind erosion and break the impact of raindrop splash, slowing down the velocity of surface runoff and impeding water erosion. Reduced runoff results in a reduced loss of water, soil, fertilizer and pesticides, so avoiding wastes and contamination of soil and downstream waters. Crop residues also suppress weed growth.

— The soil cover also makes the soil organic matter content to augment over time, increasing soil fertility and improving the structure. Soil organic matter binds the soil particles together into structural units called aggregates and thus helps to maintain a loose, open, granular soil structure. Such a friable soil structure improves water infiltration, retention and availability, impedes water runoff and thereby soil erosion. In turn, improved water infiltration and the reduction of moisture loss by evaporation improve the capacity of the soil to retain nutrients and moisture.

— Crop residues are also a habitat and a source of food for the organisms in the soil, which in turn help the formation of stable aggregates. Stimulation of the biological activity in the soil (micro-organisms and insects) and in the field (predators) creates conditions for effective biological pest and disease control and, in general, has a positive impact on agrobiodiversity.

— Better soil structure and increased fertility improve the rooting conditions for plant development and growth, and reduce the probability that the crops will suffer from drought and other natural disasters.

— Crop rotation and intercropping maintain and enhance soil fertility, while crop rotation helps to break insect pest, disease and weed cycles. The inclusion of leguminous green-manure or cover crops in small-farm systems not only provides dense cover and large quantities of organic matter to the soil, but also significant quantities of microbially fixed nitrogen.

Conservation agriculture also contributes to wider environmental benefits such as:

— Less erosion impedes land degradation and desertification.

— Reduced runoff limits the loss of water and soil, but also of fertilizer and pesticides, and so avoids the contamination of soil as well as pollution and siltation of downstream waters.

— No-tillage and mulching reduce the release of carbon into the atmosphere. And sequesters more carbon which mitigates climate changes and the impact of greenhouse gases.

— Biodiversity above and below the ground is enhanced through diversification, improved field conditions and stimulation of biological activity (soil micro-organisms but also pest predators). Living cover crops and crop residues provide the habitats for a variety of animals (insects, birds, small mammals, reptiles, and so on), plants and micro-organisms, which are necessary to sustain key functions of the agro-ecosystem.

The environmental impacts of CA depend critically on whether and to what extent the basic conservation principles are applied as well as the length of time they have been practised. In order to assess the fertility status of the fields, and thereby compare the extent of soil degradation, under the survey 71 soil samples from CA fields and 51 from adjacent non-CA fields were tested for nutrient analysis. According to the soil analysis, the nutrients and the organic matter contents are higher in the likoti fields than in the ploughed ones, but the difference is not statistically significant. This small difference may depend on the fact that, at the time sample were tested, the households had being practising likoti for two to four seasons, and it was thus too early to expect any significant change in the fields' soil health status. Nonetheless, according to a soil fertility index (SFI) built to compare CA with conventional fields, the overall soil fertility was higher in the former. This finding suggests that in the CA fields soil fertility was building up or at least was not degrading as fast as in the ploughed fields.

This analysis, although partial, largely confirms the results achieved by most of the experiential literature on conservation agriculture. However, in order to fully assess the impacts of likoti on soil and water conservation further research should be conducted on a long-term basis.

The Role of Social Capital

The positive impact of social capital on the effective use of SWC measures has been demonstrated by several empirical studies in Uganda, Kenya, Philippines and Peru. Indeed, several social capital dimensions affect the relevance of the factors determining the adoption of CA. For instance, higher levels of trust and reciprocity help farmers to access labour and credit (through labour exchanges, social networks and associations), thus reducing the need for external incentives. By fostering cooperation and collective action, social capital also facilitates extension and field activities, and encourages adaptive research by enabling the formation of farmer groups and networks among researchers, extensionists and farmers at different levels. As a means to support institutional agreements, avoid conflicts and foster community participation, social capital may also help to solve the problems related to the use of common pool resources, such as land tenure and grazing rights, which seriously affect the adoption of CA in SSA. The presence of social capital also supports a receptive attitude towards the cultural and institutional changes that accompany technical transformations in any process of innovation adoption and diffusion.

According to the information collected from key stakeholders, the socio-economic suitability of likoti is in fact counterbalanced by a number of cultural and relational issues. For example, since the ox-plough is identified with a particular social status, some Basotho stigmatize the practice due to the fact that labour is provided by people instead of animals. Also, the customary rules which allow villagers to collect the crop residues and herd their livestock in the fields after the harvest may represent an obstacle to the adoption of CA. Farmers who do not allow the neighbours into their fields in order to keep the mulch cover and avoid soil compaction, could incur relational social problems with the rest of the community. In order to assess how these aspects affect the socio-economic sustainability of CA in Lesotho, the following paragraphs analyse the role of social capital and other culture related issues in the process of adoption and diffusion of likoti.

On the basis of the data collected in both phases of the field survey, ten social capital variables were established. Four sought to evaluate the rate of participation in formal and informal networks or the "structural" dimensions of social capital. Six variables were related to the cognitive aspects of social capital, such as trust and reciprocity. The different forms of social capital which characterize the sample were then analysed along with the socio-economic and farming related variables, in order to assess possible relationships and dependencies. The empirical analysis was conducted with the support of Bayesian networks, whose structure—learnt inferentially from the data—reflect the (conditional) dependencies among the variables.

From the analysis it emerged that the households who adopted likoti are more endowed with social assets than those who did not. In particular, a "network dimension" characterizes CA farmers in the lowland sites, while a "trust dimension" is stronger among respondents in the mountains. These differences somehow reflect the different impacts that the socio-economic trends discussed earlier are having on the local communities. In the lowlands, where temporary migration to South Africa and urban and peri-urban areas is more frequent, community and kinship linkages deteriorate faster than in the mountains and traditional coping strategies are less effective. At the same time, "looser", choice-based networks with balanced reciprocity, are substituting community groups that used to rely on generalized reciprocity. Even though these networks represent an important asset for their members, compared to the past, vulnerable groups are more likely to be marginalized. In the mountain sites, instead, the "trust dimension" seems to be closely related to the persistence of traditional institutions, including community support mechanisms. These findings are supported also by the information provided by key interviewees. For instance, all the chiefs reported that, mostly due to the general impoverishment of the people, the quality of relationships and the degree of cooperation are worse than in the past. However, interviewees in the lowlands especially stressed the decrease in reciprocal trust and the diffusion of a nascent "payback" mentality.

Apart from these location specificities, the Bayesian networks learnt from the social capital variables show that the two dimensions of social capital—network and trust—are interrelated, and that both are linked to the "adoption" variables, suggesting that a higher endowment with social assets, as in the case of CA farmers, has fostered the adoption of the innovative practice. The structural learning of Bayesian networks also found that the degree of knowledge on CA is strongly correlated to the attainment of training, and the effect of training on the degree of knowledge is stronger in the mountains than in the lowlands. That is, likoti adopters in the mountains—who revealed to be more cooperative—have a better knowledge of conservation agriculture principles and apply them more correctly than adopters in the lowlands do. Most likely, these differences depend on the different approach used by

CA trainers in Qacha's Nek (the mountain district) and in Butha-Buthe and Leribe (in the lowlands). As mentioned earlier, in fact, the former interact frequently with the trainees, organize field visits and gatherings to discuss problems and issues, and also encourage farmers to work collectively. On the contrary, in the lowlands, trainers complain that they cannot rely on farmers' cooperation in spite of their efforts to foster it. Two important conclusions can be drawn from the above discussion:

— Social capital is an important determinant of the adoption of CA practices. However, the higher level of trust among respondents along with the participatory approach used by CA trainers, have been especially relevant to both the performance and the acceptance of the technology. These findings confirm the critical role that a proper combination of social capital and capable agency of committed leaders play in the achievement of local development objectives, as highlighted in the recent literature.

— The conservation farming practice revealed to be very suitable to the local socio-economic conditions, as well as to the environmental constraints. However, the long-term sustainable impacts on crop intensification as a means to sustain livelihoods depend greatly on social, cultural and institutional factors.

CA and Sustainable Crop Production Intensification

The rationale for applying planting basins in Lesotho is threefold: to stop and reverse land degradation, especially soil erosion, to obtain higher yields and to improve food security. This section briefly summarizes the results of the analysis. It then focuses on the main lessons learnt about the factors that so far have mostly determined the adoption and the performance of CA in Lesotho. It concludes with a discussion about the areas for potential improvement.

Outcomes of the Activities

The analysis of the survey data has shown that the adoption of likoti had brought about significant advantages compared to conventional tillage practices. The most important are:

— Higher agricultural productivity, due to improved efficiency in the use of inputs and other resources.

— Higher social sustainability, due to the accessibility to the technology by all social categories, including the most vulnerable.

— Greater environmental sustainability, due to improved soil structure, enhanced fertility and reduced erosion

Farmers using likoti have achieved significantly larger yields by employing relatively fewer means. The economic analysis and the assessment of returns to labour suggest that the new CA practice is profitable notwithstanding the significant workload needed especially in the first two seasons in setting up the likoti. Furthermore, if well managed, the workload necessary to prepare the land can be spread over the dry season and over several seasons, thereby relocating the heavy labour out of the peak planting period and also spreading it over time. Preparing the field during the dry season also enables farmers to sow earlier and benefit from timely planting as well as to programme other off-farm activities. In any case, the overall amount of net labour required tends to decrease over time. After the first season, in fact, it is not necessary to design the grid and break the hardpan to dig the basins. In addition, it has been demonstrated that by weeding CA fields frequently in early years and at the right time, the weed infestation reduces progressively over time thus making weeds easier and easier to control. Nonetheless, the availability of labour has to be carefully assessed just because timely and precise farming management is critical to the achievement of positive results.

The spread of conservation agriculture has a great potential to improve the degree of agricultural knowledge, especially with regard to the management of the agro-ecosystem, also among the farmers who continue to use conventional tillage methods. In fact, beyond introducing innovative farming practices, it has increased people awareness about the causes of land degradation and related possible solutions. On the other hand, the likoti farmers themselves sometimes fail to apply strictly all the CA principles, as recommended by the promoters of the practice. In some cases, this may depend on the scarcity of the resources available (for instance, this can justify low applications of fertilizer or infrequent weeding); in other cases, cultural related issues hamper the correct adoption (e.g., farmers can not maintain an adequate soil cover because other villagers are allowed to collect the stoves for fuel or fodder). As a result, the potential of CA in Lesotho is still underexploited.

Agriculture in Lesotho is mainly subsistence oriented, but still plays an important role also as complementary source of income. The adverse climate conditions and the fragile environment, along with the progressive impoverishment of the population, have caused agricultural yields to steadily decrease over the last years. At the same time, social and cultural breakdowns have weakened people's ability to farm, which depends heavily on social assets and relational skills. Likoti has shown meaningful potential to improve livelihoods and food security employing fewer inputs compared to conventional farming practices. Indeed, the suitability to the local socio-economic conditions—that is, the accessibility to the technique by all social categories—is even more important than the impact on yields (still very significant).

The spread of the practice is also having significant impacts on the environment. Improved fertility and a stronger soil structure are the primary causes of the increase in yields. At the same time, they contribute to stop and reverse the process of soil erosion and land degradation which dramatically affects the Lesotho landscape. The social and the environmental sustainability associated to the CA practice are extremely important. On the one hand, poor resource farmers obtain higher yields, improve household food security and possibly, in the longer run, they will be able to make a living from farming. On the other hand, the positive social and environmental impacts can contribute, on a larger scale, to a sustainable development process based on the revitalization of the agricultural sector and the preservation of the natural resource base.

Lessons Learnt

According to the analysis, the factors that so far have mostly determined the adoption of CA in Lesotho are:

— *Economic incentives*: Lack of assets and income, as well as the socio-demographic features of the farmers and their households, do not affect the possibility to adopt the technology. However, in order to start practising CA, vulnerable households may need some support for buying the inputs and be sure to get enough yields so that they can continue to farm with their own resources. Indeed, the majority of the respondents received some forms of subsidy—either inputs or food—to start likoti but this support, except in a few cases, stopped after the first season. Furthermore, when asked about the main reason for starting CA, only 5% of the farmers who had already abandoned CA and 8% of those who were still practicing mentioned "the provision of food and other incentives." This means that the wide majority of the farmers would continue to use likoti even after subsides have stopped and would confirm the economic sustainability of the practice.

— *Human capital*: CA farmers are more educated than conventional farmers. In particular, female adopters, who are the less endowed with economic assets and other resources, are significantly more educated (even though at low levels of education) than the other categories, and especially compared to female conventional farmers. Thus, human capital has been found to be an important determinant of adoption, especially when access to other resources is limited, as it is the case for many Basotho women.

— *Social capital:* CA farmers are more endowed with social assets than conventional farmers. In particular, a "network dimension" characterizes CA farmers in the lowlands, while a "trust dimension" is stronger among respondents in the mountains. The two dimensions are interrelated, and both seem to affect the decision to adopt CA. These findings are consistent with the most recent literature on agricultural innovation which highlights the importance of social capital in the adoption and the performance of innovative agricultural practices. In the Lesotho case, higher degree of trust and cooperation among community members may help to find institutional agreements and organizational solutions to overcome some of the constraints to the correct application of the conservation principles, for instance issues related to land tenure and the integration of farming and livestock activities.

— *Training and capable agency*: Knowledge on CA is found to be strongly correlated to the attainment of training, and the effect of training on the degree of knowledge imparted is stronger in the mountains than in the lowlands. The finding may be well explained by the participatory approach used by the trainers in the mountains, combined with the farmers' good attitude towards cooperation and trustfulness. The positive impact of the commitment of CA trainers in Qacha's Nek on the performance and the acceptance of the technology, confirms the critical role that a proper combination of social capital and capable agency play in the achievement of local development objectives, as highlighted by the literature.

Drawing on the discussion above, a number of consistent policy implications for the successful diffusion of innovative conservation practices, not only in Lesotho but also in other Sub-Saharan countries can be derived. The most important are:

Incentives to the adoption of CA practices

The critical impact that a proper application of conservation practices can have on the livelihoods of vulnerable household categories justifies the provision of some incentives, such as inputs or small loans, that would enable them to start practising. Food aid and other forms of subsidies, although useful for those households who need to recover their livelihood basis, should be used carefully in order not to create dependency or discourage food production. In all cases, support should be given on a provisional basis and under payback schemes. Moreover, participation in training and demonstration activities should be a necessary condition to receive further assistance.

Initial labour intensiveness is still a major deterrent to the adoption of likoti, just as happens in other African countries where similar planting basin systems have been adopted. If the fields are properly managed, the net workload decreases over time, but if there is not enough workforce available, it becomes harder and harder to harness the potential benefits. Thus some kind of temporary support may be necessary also to overcome the possible initial labour constraint. For example, programmes targeting vulnerable households can employ landless workers on others' fields.

Finally, beyond having positive impacts on agricultural yields and food security, CA has also a critical role in the conservation of the environment and the natural resources. The environmental impacts can be considered as positive externalities from which the whole society benefits, but they are not perceived by individuals, especially when adequate policy support is lacking. Furthermore, most of the social benefits (and also some private ones) manifest themselves in the medium to the longer term, whereas the highest costs have to be sustained in the early adoption phases. Therefore, public-funded support to the spread of CA may be justified also by the need to overcome this dichotomy between the social desirability of the technology and its on-farm attractiveness. To this aim, public support should include not just economic subsides and other forms of direct incentives, but also more effective advisory services and information campaigns.

Education, Information and Advocacy

More information about the concept of conservation agriculture and its potential advantages should circulate countrywide, not just among potential adopters. The main objectives of broad information campaigns should be to reduce scepticism and achieve a wider acceptance of CA also amongst the conventional farmers and to raise awareness about the long-term environmental and social benefits. The 'supply' of information should be accompanied by an adequate investment in farmers' receptive capacities. In other words, whenever necessary, a more general effort is needed to enhance education, which indeed was found to be particularly relevant to women's involvement.

Special training and information sessions have to be conceived for local researchers, officials of the Ministry of

Agriculture and extension staff in order to provide farmers with both training and technical assistance. In particular, a deeper involvement of the extension staff in training and field activities would also foster a wider acceptance of the CA practices.

Farmers' Participation and Training

The importance of participatory processes in the adoption of agricultural innovation has been extensively discussed in the literature. Participation at community level of all members, and especially of the local leadership, allows a better understanding and a wider acceptance of new ideas and practices, especially if they need not just a technical shift but also a radical cultural change, as in the case of conservation practices. The survey has identified a number of issues that community would better address through a more participatory approach. For instance, access by herders and other villagers into CA fields after the harvest is one of the most important deterrents to the correct application of CA principles. In order to overcome it, community members should not just discuss the issue of herding livestock out of the fields. They should also find feasible community-based solutions for the livestock owners and alternative fodder and fuel sources. Similarly, a closer integration of farming and livestock could help overcome other constraints, such as the supply of organic manure, or the production of fodder crops in rotation with other crops. Under such approach, common rules on range lands and promotion of CA would be complementary elements of an integrated strategy that aim to combat land degradation and conserve the natural resource base.

The extent of farmers' participation is also important with regard to training. The effectiveness of CA practices largely depends on the timely and appropriate management of all the farming activities. Therefore, the enhancement of technical knowledge and precision skills through adequate training is critical. Equally important is the approach used by the trainers. It has been demonstrated, in fact, that the promotion of participatory field activities and a close interaction between farmers and trainers lead to the better assimilation of CA principles and, in turn, to a more appropriate application. However, this formula is more likely to succeed where the degree of trust and cooperation among farmers, i.e. social capital, is higher.

Areas For Further Research And Improvement

The diffusion of likoti has brought about several advantages to its adopters, but its potential is still underexploited. As already stressed, in fact, not all farmers are strictly following the CA principles, and this may depend on technical and/or resource constraints but also on cultural bias and institutional problems. The previous section discussed a number of possible actions to be taken to further enhance the positive impacts of CA. This section focuses on those factors that, albeit commonly identified as important determinant of the adoption of CA, are still lacking or absent in Lesotho.

Among the most important, effective policy support has been virtually absent. In spite of Government's acknowledgement of the benefits associated with minimum tillage techniques, the concrete involvement of the Ministry of Agriculture and Food Security (MOAFS) in the diffusion of CA has been limited. The low commitment of the MOAFS is reflected also in the lack of support from the extension services, as admitted also by the local extension officers interviewed during the field survey. At policy level, neither the creation of the Conservation Farming Network Group (CFNG) nor the support that FAO and WFP gave to the organizations that first promoted likoti, translated in a functioning multiple stakeholder partnership. Lack of interaction among farmers and other actors, including extension services, in turn affected the degree of farmer participation in the diffusion and, most importantly, in the adaptation of the technology to the local conditions.

The low interaction of formal research and farmers' indigenous knowledge has been another important missing aspect. In particular, institutional issues related to land tenure, such as the use of stubble as fuel or fodder by the villagers, and the integration of the livestock and the farming systems, have received inadequate attention by the promoters of CA. In order to properly adapt the conservation principles to the local agro-ecological conditions, as well as to the cultural beliefs, it would be very helpful also to assess the innovations developed by the farmers themselves. However, neither the NGOs and the international organizations involved nor the MOAFS have significantly interacted with innovating farmers.

Borrowing the lexicon from the Agricultural Innovation System (AIS) approach, in Lesotho the promotion of innovative CA practices has not been based on "a dynamic process of interacting embedded in specific institutional and policies contexts". Limited stakeholder interaction and inadequate policy support not only limit the potential benefits associated with the use of the technique, but they also hamper the internalization of social costs and benefits, discourage the social acceptance of the innovative practices, and ultimately affect the rate of adoption.

Adaptive research based on constant interaction among formal researchers, technology promoters and local farmers is especially important in the diffusion of CA practices, just because of their flexible nature. In order to fully exploit the benefits of a technology that can be suited to different environment conditions, in fact, farmers need to enhance their innovation capabilities, which are not only of a technical nature. Participatory research activities are critical also in order to include aspects of indigenous knowledge and traditional institutions, and ultimately facilitate the tremendous mind shift that has to take place in the transition from conventional to conservation farming practices and which is one of the biggest challenges to the adoption of CA. However, according to Fowler and Rockstrom, "identification and recognition of local traditions or indigenous knowledge is important, but it is the actual possibility of building on these that has real potential. [...] For this approach to succeed, the social environment should be conducive, the intervention must involve communities not individuals, the activities must involve all potential players."

A critical component of a working participatory and adaptive research system should thus be the creation of multiple stakeholder partnerships. As the Swiss Commission for Research Partnerships with Developing Countries (KFPE) has put it, "ideally, a research partnership should strive for a dynamic equilibrium in which all involved parties are open to a multiple transformation in terms of mutual learning, cultural understanding, scientific upgrading, capacity building, and attitudinal behaviour towards all partners. Applying trans-disciplinary or multilevel, multi-stakeholder approaches, where all relevant stakeholders are actively participating, helps generate meaningful results and fosters processes that promote impact. In such partnerships all partners have a voice in decision-making processes and their capacities are used and further developed in a complementary and most fruitful way".

Very recently, in July 2009, Growing Nations—Rev. Basson led organization—signed an agreement with the International Fund for Agricultural Development (IFAD) for the provision of a grant to establish a training centre and to develop new training curricula. A key step will be to collect the training material that already exists and give it to farmers in a workable format. Further challenge will be the establishment of intensive training modules for agricultural students. For this reason, Rev. August Basson decided to move Growing Nations' activities to Maphutseng, in Mohale's Hoek district, where the centre, and the possibility to learn likoti, is more easily accessible for farmers from the whole country. Playing an adequate catalyst role, this initiative could provide the right incentive for more actors to get involved in the diffusion of the practice. At the same time, the centre is in a privileged position to start conduct systematic research on this and other conservation practices, following an adaptive, participatory approach.

Sustainable crop production intensification should contribute to the economic development of farming communities and seek to stimulate local economies through understanding and development of local, national and regional markets.

It is relevant to mention that for two successive years, in 2007 and 2008, some likoti farmers who before adopting CA were unable to produce enough maize for themselves, were able to sell excess grain production to the World Food Programme (WFP). The case confirms the concrete possibility for likoti farmers to shift from subsistence to commercial agriculture. However, at the current status, most CA farmers are still far from achieving this goal. Furthermore, to boost production is just one of the leverages needed to develop the commercial agricultural sector. Adequate policies—and related investments—should also address many other factors: the provision of adequate infrastructure, the strengthening of market linkages, product differentiation into niche and high value markets, among the others.

This section has pointed out some issues that would be worthy of more careful assessment, both in research activities and at policy level. In spite of these areas for improvement, however, the results of the analysis have shown that the diffusion of likoti has already led to sustainable crop production intensification by enhancing soil fertility and consequently crop yields. These results are especially significant if they are put in the context of growing vulnerability which has been characterizing Lesotho in recent years. Indeed, the suitability of the conservation agriculture practice to different social and economic conditions, even the poorest, is one of the most important benefits associated with its adoption. As a farmer well put it, the main advantage of likoti is just that "Everybody can do it".

Adoption of CA in SSA: Constraints and Opportunities

Associated with decreasing variable costs, larger outputs increase net profitability and help to strengthen and diversify rural livelihoods. Further important benefit is the enhancement of food security, due to larger outputs as well as to nutritional improvements. A more diversified diet results from the availability of diverse crops planted in rotation or along with cereals, the extension of cultivated spaces (thanks to additional resources, including time), and—on a longer term—increased income.

Beyond increasing net profitability and food security, CA ensures long-term socio-economic and environmental sustainability, even in densely populated areas, such as occurred in Burkina Faso. African soils are increasingly affected by land degradation and desertification, but these phenomena have to be contextualized in order to understand the local-specific causes, features and consequences. CA may benefit African agriculture just because of the possibility to adapt a wide range of practices and techniques to different socio-economic contexts as well as to different agro-ecological conditions. In semi-arid lands, conservation agriculture retains water and moisture in the soil, keeps the soil temperature even, and protects the land from erosion during heavy downpours. In sub-humid and humid areas, crops planted at closer spaces and cover crops help suppress weeds and protect the soil. On slopes, conservation agriculture reduces runoff and soil erosion, and can be effectively used in association with terraces, contour grass strips and other erosion control methods.

Further advantages for African farmers stem from lower labour requirements. In many countries, the rural population is steadily being reduced by rural-urban migration and by the HIV/AIDS pandemic. These phenomena concern particularly the younger male population, meaning that those with the best potential for heavy physical work are no longer working on the land, while a growing number of households are headed by women. At the small scale, once

the conservation farming system is well established, the shorter time required for land preparation and weeding, along with the more even distribution of labour throughout the year, allow to reduce the workload during the peak season. Also draught animal power (DAP) systems permit labour saving by shifting from the mouldboard plough to shallow ripping. Time saving and reduced drudgery of field activities benefit all farmers, but a major opportunity arises for women.

On the other hand, the adoption of CA may in some cases imply higher labour requirements. Especially in the transition phase, many small-scale conservation farming practices require temporary additional work to prepare the land and for weed and pest control. In particular, land preparation by digging planting pits and frequent weeding may impose an extremely hard and time consuming extra burden on women. However, to assess this kind of impacts may be difficult since the division of labour among household members is not always clear-cut and it differs from place to place and from family to family.

Even though labour requirements diminish after the first or the second season (leading to the important advantages already mentioned), such initial effort may prevent resource poor farmers, and especially women and the elderly, from adopting the technology. The shift from conventional to conservation practices may also require additional financial resources to buy inputs and purchase (or hire) and maintain new equipment. The choice of an appropriate combination of CA practices and inputs should limit the expenses. However, if additional costs cannot be avoided (especially those arising from the initial labour demand), poorer, more risk-adverse farmers might need to be supported with some kind of incentives.

An important constraint to the adoption of conservation farming in Africa is represented by the difficulty of keeping a permanent soil cover due to insufficient biomass production. This may be due to agro-ecological and weather conditions, such as high humidity or poor rainfalls. But the most important reason for scarce soil cover is that available crop residues serve many other scopes such as animal feed, fencing, and fuel. For CA to be successfully adopted, substitute sources of fuel and fodder should thus be found. Otherwise, crop rotations and cover crop cultivation should allow the production of enough residues to meet the various needs. Proper crop selection, right choice of crop rotation and improved crop residues management may help achieve these objectives. For instance, many Burkinabe farmers who have started CA have then invested in livestock, since they could increase the production of fodder crops. However, especially in drier areas, sufficient biomass production still requires a lot of time and resources, and in most African countries, such as in Zambia—where the benefits of CA have been extensively described—most fields remain uncovered. Apart from technical answers, the even allocation of crop residues among multiple purposes also require the appropriate involvement of institutions, community participation and cooperation.

Another aspect that must be addressed jointly by community members is the integration of crop and livestock production, especially wherever livestock constitutes a major component of the local economy. The integration of livestock into CA may contribute significantly to the overall efficiency of the local agricultural system. Farmers can introduce forage crops into crop rotations, and these can be used for both fodder and soil cover, as well as to reduce pest problems. On the other hand, animal manure can be exploited to recycle nutrients in the fields, thereby reducing the environmental problems caused by intensive livestock production. However, conflicts between the use of organic matter to feed the animals or to cover the soil may still be difficult to solve. In many societies traditional rules allow animals to graze on stubble. Apart from reducing the biomass available for soil cover, free communal grazing on harvested fields causes soil compaction. Livestock keepers and CA farmers should find alternative solutions such as fencing animals out, planting unpalatable cover crops, enforcing land tenure arrangements, changing traditional grazing rights, or growing special plots of fodder crops. Integration of crop and animal production systems is therefore essential for sustainable rural livelihoods, not just for technical reasons but also—and in some cases especially—because of the intrinsic cultural value that farming and livestock have in African societies.

Indeed, in Africa, cultural and institutional issues affect the adoption of innovative practices, including CA, more than elsewhere. The shift to conservation practices involves a profound mindset change, and this may be especially difficult where practices such as ploughing, clearing the land

and free animal grazing are embedded in local institutions. Many African societies are also influenced by the idea that the current situation cannot be changed and that those who are born in poverty will die in poverty. In such cases people—and especially vulnerable categories, like women—are highly risk adverse and purposely avoid transformations which are likely to improve their situation on the failure of some CA experiences in South Africa). In most cases, such behaviour is rational. For instance, where women do not have rights to the land, any improvement they achieve in crop production may expose them to the risk of loosing the fields which become more attractive to their male relatives. The role of human and social capital is extremely important in overcoming constraints arising from institutional and cultural factors. Therefore, the adoption and the diffusion of CA in SSA may be held back by a lack of human and social capital, especially among the most marginalized groups and less resource endowed farmers.

This brief discussion highlighted many significant opportunities arising from the adoption of CA in SSA. At the same time, there are also some downsides mostly related to factors that may delay or corrupt the effective adoption of the practices. Some of these constraints may in turn transform into opportunities, but an appropriate set of initial conditions and/or adequate policies and interventions is needed. Based on the recent literature review, the factors that have been found to determine the successful adoption of CA in Africa can be summarized as follows:

— Incentives, usually employed to compensate for initial additional costs of technology adoption. In Zambia, Ghana and, for instance, incentive schemes have facilitated a relatively quick rate of adoption, but their long-term sustainability has not be confirmed yet. In other cases, such as in Burkina Faso, conservation practices have been adopted for a long time without external incentives.

— Adequate training, effective support from extension services, and organization of field activities (exchange visits, farmer field schools, workshops, etc.). These are critical for farmers to develop appropriate technical knowledge and management skills, as has been demonstrated in Zambia, South Africa and Ghana.

— Enhanced participation and interaction between formal research and indigenous knowledge. Many socio-economic and cultural constraints can be overcome by encouraging farmer participation in the identification of the system components best suited to their specific needs. Participation and knowledge sharing among farmers and researchers also have technical implications, since they usually support and speed up adoption and adaptation of the technology, thereby ensuring its long-term sustainability.

— *Education* is important both to overcome institutional constraints and cultural biases, and to improve farmers' management skills. For instance, Haggblade and Tembo and Chomba find that in Zambia retired school teachers, draftsmen and accountants have got better results from the employment of conservation practices. Higher educational levels enhance farmers' openness to innovation adoption and adaptation.

— *Social Capital.* As stressed also in the present case study, several structural and cognitive dimensions of social capital affect the relevance of the factors determining the adoption of conservation practices. Higher levels of trust and reciprocity, as well as easier access to labour and credit, help farmers to internalize social costs and benefits associated with the shift to CA, thus reducing the need for external incentives. By fostering cooperation and collective action, social capital also facilitates extension and field activities, and encourages adaptive research by enabling the formation of farmer groups and networks among researchers, extensionists and farmers at different levels.

— *Women's inclusion.* Depending on the practice promoted, women may find it more difficult than men to get the support they need to shift from conventional to conservation agriculture. Due to cultural biases or even to the legal system, in many African countries women have restricted access to resources (land, inputs and credit), education, training and extension services. Such limitations may seriously reduce the opportunities arising from the use of conservation practices. This is why, whenever it may be needed, incentive strategies, farmer participation activities, extension and research programmes should have a gender oriented focus.

— *Policy support* (both national and international). According to Benites et al., it is especially important for a number of specific achievements such as the involvement of the private sector (e.g., for the production of locally adapted equipment and inputs), multiple stakeholder partnerships, and the promotion of adaptive research.

3

No-tillage Farming

Since the early 1960s farmers have been urged to adopt some form of conservation tillage to save the planet's soil, to reduce the amount of fossil fuels burnt in growing food, to reduce runoff pollution of our waterways, to reduce wind erosion and air quality degradation and a host of other noble and genuine causes.

Charles Little in *Green Fields Forever* epitomized the genuine enthusiasm most conservationists have for the technique. But early farmer experience, especially with no-tillage, suggested that adopting such techniques would result in greater short-term risk of reduced seedling emergence, crop yield or, worse, crop failure, which they were being asked to accept for the long-term gains outlined above. Farmers of today were unlikely to see many short-term benefits of their conservation practices.

Leaving a legacy of better land for future generations was one thing, but the short-term reality of feeding the present generation and making a living was quite another. Not unreasonably, short-term expediency often took priority. Although some countries already produce 50% or more of their food by no-tillage (e.g. Brazil, Argentina and Paraguay), it is estimated that, worldwide, no-tillage currently accounts for only some 5–10% of food production. We still have a long way to go. Certainly there have been good, and even excellent, no-tillage crops, but there have also been failures. And it is the failures that take prime position in the minds of all but the most forward-looking or innovative farmers.

Tillage has been fundamental to crop production for centuries to clear and soften seedbeds and control weeds. So now we are changing history, not always totally omitting tillage (although that is certainly a laudable objective) but significantly altering the reasons and processes involved. Most people understand tillage to be a process of physically manipulating the soil to achieve weed control, fineness of tilth, smoothness, aeration, artificial porosity, friability and optimum moisture content so as to facilitate the subsequent sowing and covering of the seed. In the process, the undisturbed soil is cut, accelerated, impacted, inverted, squeezed, burst and thrown, in an effort to break the soil physically and bury weeds, expose their roots to drying or to physically destroy them by cutting. The objective of tillage is to create a weed-free, smooth, friable soil material through which relatively unsophisticated seed drill openers can travel freely.

During no-tillage, few, if any, of the processes listed above take place. Under no-tillage, other weed-control measures, e.g. chemicals, must substitute for the physical disturbance during tillage to dislodge, bury or expose existing weeds. But part of the tillage objective is also to stimulate new weed seed germination so that fresh weeds get an 'even start' and can therefore be easily killed in their juvenile stages by a single subsequent tillage operation. No-tillage, therefore, must either find another way of stimulating an 'even start' for new weeds, which would then require a subsequent application of herbicide or avoid stimulating new weed growth in the first place.

In his keynote address to the 1994 World Congress of Soil Science, Nobel Prize-winner Norman Borlaug estimated that world cereal production (which accounts for 69% of world food supply) would need to be raised by 24% by the year 2000 and doubled by the year 2025. More importantly, Borlaug estimated that grain yields would need to increase by 80% over the same time span because creating new arable land is severely limited throughout the world. Until now, yield increases have come largely from increased fertilizer and pesticide use and genetic improvement to the species grown. The challenge is for no-tillage to contribute to future increases, while simultaneously achieving resource preservation and environmental goals. But this is only going to happen if no-tillage is practised at advanced technology levels.

The notion of sowing seeds into untilled soils is very old. The ancient Egyptians practised it by creating a hole in untilled soil with a stick, dropping seeds into the hole and then closing it again by pressing the sides together with their feet. But it was not until the 1960s, when the herbicides paraquat and diquat were released by the then Imperial Chemical Industries Ltd (now Syngenta) in England, that the modern concept of no-tillage was born because now weeds could be effectively controlled without tillage.

For the preceding decade it had been recognized that, for no-tillage to be viable, weeds had to be controlled by some other method than tillage. But the range of agricultural chemicals then available was limited because of their residual effects in the soil. A delay of several weeks was necessary after spraying before the new crop could be safely sown, which partly negated saving of time, one of the more noteworthy advantages of no-tillage compared with tillage. Paraquat and diquat are almost instantly deactivated upon contact with soil. When sprayed onto susceptible living weeds, the soil beneath is almost instantly ready to accept new seeds, without the risk of injury. This breakthrough in chemical weed control spawned the birth of true no-tillage.

Since then, there have been other broader-spectrum translocated non-residual chemicals, such as glyphosate, which was first introduced as Roundup by Monsanto. Other generic compounds, such as glyphosate trimesium (Touchdown) and glufosinate ammonium (Buster), were later marketed by other companies, which have expanded the concept even further.

In other circumstances non-chemical weed control measures have been used. These include flame weeding, steam weeding, knife rolling and mechanical hand weeding. None of the alternative measures has yet proved as effective as spraying with a translocated non-residual herbicide. These chemicals are translocated to the roots of the plant thereby affecting a total kill of the plant. Killing the aerial parts alone often allows regeneration of non-affected plant parts.

The application of any chemicals within agricultural food production correctly raises the question of human and biological safety. Indeed, many chemicals must be very carefully applied under very specific conditions for specific results, just like any of the modern pharmaceuticals that assist in cures and controls. Through careful science, and perhaps some good fortune, glyphosate has been found to be non-toxic to any biological species other than green plants and has been safely used for many years with virtually no known effects other than the control of undesired plants.

An even more recent development using genetic modification of the crops themselves has made selected plant varieties immune to very specific herbicides such as glyphosate. This unique trait permits planting the crop without weed concerns until the crop is well established and then spraying both the crop and the weeds with a single pass. The susceptible weeds are eliminated and the immune crop thrives, making a full canopy that competes with any subsequent weed growth, usually through to harvest. Only selected crops such as maize and soybean are currently commonly used in this fashion, but they have already attained a very significant percentage of the world's acreage. With this success, other important food and fibre crops are being modified for this capability.

WHAT IS NO-TILLAGE?

As soon as the modern concept of no-tillage based on non-residual herbicides was recognized, everyone, it seems, invented a new name to describe the process. 'No-tillage', 'direct drilling' or 'direct seeding' are all terms describing the sowing of seeds into soil that has not been previously tilled in any way to form a 'seedbed'. 'Direct drilling' was the first term used, mainly in England, where the modern concept of the technique originated in the 1960s. The term 'no-tillage' began in North America soon after, but there has been recent support for the term 'direct seeding' because of the apparent ambiguity that a negative word like 'no' causes when it is used to describe a positive process. The terms are used synonymously in most parts of the world, as we do in this book.

Some of these names are listed below with their rationales, some only for historical interest. After all, it's the process, not the name, that's important.

— *Chemical fallow*, or *chem-fallow*, describes a field currently not cropped in which the weeds have been suppressed by chemical means.

— *Chemical ploughing* attempted to indicate that the weed control function usually attributed to ploughing was being done by chemicals. The anti-chemical lobby soon de-popularized such a restrictive name, which is little used today.

— *Conservation tillage* and *conservation agriculture* are the collective umbrella terms commonly given to no-tillage, minimum tillage and/or ridge tillage, to denote that the inclusive practices have a conservation goal of some nature. Usually, the retention of at least 30% ground cover by residues after seeding characterizes the lower limit of classification for conservation tillage or conservation agriculture, but other conservation objectives include conservation of money, labour, time, fuel, earthworms, soil water, soil structure and nutrients. Thus, residue levels alone do not adequately describe all conservation tillage or conservation agricultural practices and benefits.

— *Disc-drilling* reflects the early perception that no-tillage or direct drilling could only be achieved with disc drills (a perception that proved to be erroneous); thus some started referring to the practice as disc-drilling. Fortunately the term has not persisted. Besides, disc drills are also used in tilled soils.

— *Drillage* was a play on words that suggested that under no-tillage the seed drill was in fact tilling the soil and drilling the seed at the same time. It is not commonly used.

— *Minimum tillage*, *min-till* and *reduced tillage* all describe the practice of restricting the amount of general tillage of the soil to the minimum possible to establish a new crop and/or effect weed control or fertilization. The practice lies somewhere between no-tillage and conventional tillage. Modern practice emphasizes the amount of surface residue retention as an important aim of minimum or reduced tillage.

— *No-till* is a shortening of no-tillage and is not encouraged by purists, for grammatical reasons.

— *Residue farming* describes conservation tillage practices in which residue retention is the primary objective, even though many of the 'conservation tillage' benefits previously mentioned may also accrue.

— *Ridge tillage*, or *ridge-till*, describes the practice of forming ridges from tilled soil into which widely spaced row crops are drilled. Such ridges may remain in place for several seasons while successive crops are no-tilled into the ridges, or they might be re-formed annually.

— *Sod-seeding*, *undersowing*, *oversowing*, *overdrilling* and *underdrilling* all refer to the specific no-tillage practice of drilling new pasture seeds into existing pasture swards, collectively referred to as pasture renovation. The correct use of the term oversowing does not involve drilling at all, but rather is the broadcasting of seed on to the surface of the ground. Each of the other listed terms involves drilling of the seed.

— *Stale seedbed* describes an untilled seedbed that has undergone a period of fallow, usually (but not exclusively) with periodic chemical weed control.

— *Strip tillage*, or *zone tillage*, refers to the practice of tilling a narrow strip ahead of (or with) the drill openers, so the seed is sown into a strip of tilled soil but the soil between the sown rows remains undisturbed. 'Strip tillage' also refers to the general tilling of much wider strips of land (100 or more metres wide) on the contour, separated by wide fallowed strips, as an erosion-control measure based on tillage.

— *Sustainable farming* is the end product of applying no-tillage practices continuously. Continuous cropping based on tillage is now considered to be unsustainable because of resource degradation and farming inefficiencies, while continuous cropping based on no-tillage is much more likely to be sustainable on a long-term basis under most agricultural conditions. Some discussions of 'sustainability' include broader considerations beyond the preservation of natural resources and food production, such as economics, energy and quality of life.

— *Zero-tillage* was synonymous with no-tillage and is still used to a limited extent today.

The most commonly identified feature of no-tillage is that as much as possible of the surface residue from the previous crop is left intact on the surface of the ground, whether this be the flattened or standing stubble of an arable crop that has been harvested or a sprayed dense sward of grass. In the USA, where the broad category of conservation tillage is generally practised as an erosion-control measure, the accepted minimum amount of surface covered by residue after passage of the drill is 30%. Most practitioners of the more demanding option of no-tillage or direct seeding aim for residue-coverage levels of at least 70%.

Of course, some crops, such as cotton, soybean and lupin, leave so little residue after harvest that less than 70% of the ground is likely to be covered by residue even before drilling. Such a soil, however, can be equally well direct drilled as a fully residue-covered soil in the course of establishing the next crop. Thus it is also regarded as true no-tillage. What is no-tillage to one observer may not be no-tillage to another, depending upon the terms of reference and expectations of each observer.

The most fundamental criterion common to all no-tillage is not the amount of residue remaining on the soil after drilling, but whether or not that soil has been disturbed in any way prior to drilling. Even then, during drilling, as will be explained later, such a seemingly unambiguous definition becomes confused when you consider the actions of different drills and openers in the soil. Some literally till a strip as they go, while others leave all of the soil almost undisturbed. So the untilled soil prior to drilling might well become something quite different after drilling.

This book is focused on the subject of 'no-tillage' in which no prior disturbance or manipulation of the soil has occurred other than possibly minimal disturbance by operations such as shallow weed control, fertilization or loosening of subsurface compacted layers. Such objectives are entirely compatible with true no-tillage. Any disturbance before seeding is expected to have had very minimal surface disturbance of soil or residues.

Depending on the field cropping history and the available seeding machine capability, it may be necessary to perform one or more very minimal-disturbance functions for best crop performance. The most common of these needs is the application of fertilizer when that function can not be made part of the seeding operation. Early no-tillage seeding trials often simply broadcast the fertilizer over the soil surface expecting it to be carried into the soil profile by precipitation, but two things became readily apparent. First, only the nitrogen component was moved by water, leaving the remaining forms, such as phosphorus and potassium, on or near the soil surface. And even then preferential flow of soluble nitrogen down earthworm and old root channels often meant that much of it bypassed the juvenile roots of the newly sown crop

Secondly, emerging weeds between the crop plants readily helped themselves as the first consumers of this fertilizer and 'outgrew' the crop. Subsurface placement is now the only recommended procedure, often banded near the seeding furrow or emerging crop row.

Where herbicides are less available, it may prove more economical to perform a weeding pass prior to seeding to reduce the weed pressures on the emerging crop. If used in conservation agriculture, this operation must be very shallow and leave the soil surface and residues nearly intact ready for the seeding operation. Typical implements that can achieve this quality of weed control are shallow-running V-shaped chisels or careful hand hoeing.

Historical compaction arising from many years of repetitive tillage often cannot be undone 'overnight' by switching to no-tillage. While soil microbes are rebuilding their numbers and improving soil structure, a process that may take several years even in the most favourable of climates, historical compaction may still exist. Temporary relief can often be achieved by using a subsoiling machine that cracks and bursts subsurface zones while causing only minor disturbance at the surface.

But sometimes overly aggressive sub-soilers cause so much surface disturbance that full tillage is then required to smooth the surface again. This seemingly endless negative spiral must be broken if the benefits of no-tillage are to be gained. All that is required is a less aggressive or shallow-acting subsoiler that allows no-tillage to take place after its passage without any further 'working' of the soil surface layer.

Another effective method is to sow a grass or pasture species in the compacted field and either graze this with light stocking or leave it ungrazed as a 'set-aside' area for a number of years before embarking on a no-tillage programme thereafter without tillage. A rule of thumb for how many years of pasture are required to restore soil organic carbon (SOC) and ultimately the structural damage done by tillage was established by Shepherd *et al.* for a gley soil (Kairanga silty clay loam) under maize in New Zealand soils as:

— Where tillage has been undertaken for up to 4 consecutive years, it takes approximately 11 years of pasture to restore SOC levels for each year of tillage.

— Where tillage has been undertaken for more than 4 consecutive years, it takes up to 3 years of pasture to restore SOC levels for each year of tillage.

The rate of recovery of soil structure lags behind the recovery rate of SOC. The more degraded the soil, the greater the lag time.

Why No-tillage?

It is not the purpose of this book to explore in detail the advantages and disadvantages of either no-tillage or conservation tillage. Numerous authors have undertaken this task since Edward Faulkner and Alsiter Bevin questioned the wisdom of ploughing in *Ploughman's Folly* and *The Awakening*. Although neither of these authors actually advocated no-tillage, it is interesting to note that Faulkner made the now prophetic observation that 'no one has ever advanced a scientific reason for ploughing'. In fact, long before Faulkner's and Bevin's time, the ancient Peruvians, Scots, North American Indians and Pacific Polynesians are all reported to have practised a form of conservation tillage.

None the less, to realistically focus on the methods and mechanization of no-tillage technologies, it is useful to compare the advantages and disadvantages of the technique in general as measured against commonly practised tillage farming. The more common of these are summarized below with no particular order or priority. Those followed by an " can be either an advantage or a disadvantage in differing circumstances.

Advantages

— *Fuel conservation.* Up to 80% of fuel used to establish a crop is conserved by converting from tillage to no-tillage.

— *Time conservation.* The one to three trips over a field with no-tillage (spraying, drilling and perhaps subsoiling) results in a huge saving in time to establish a crop compared with the five to ten trips for tillage plus fallow periods during the tillage process.

— *Labour conservation.* Up to 60% fewer person-hours are used per hectare compared with tillage.

— *Time flexibility.* No-tillage allows late decisions to be made about growing crops in a given field and/or season.

— *Increased soil organic matter.* By leaving the previous crop residues on the soil surface to decay, soil organic matter near the surface is increased, which in turn provides food for the soil microbes that are the builders of soil structure. Tillage oxidizes organic matter, resulting in a cumulative reduction, often more than is gained from incorporation.

— *Increased soil nitrogen.* All tillage mineralizes soil nitrogen, which may provide a short-term boost to plant growth, but such nitrogen is 'mined' from the soil organic matter, further reducing total soil organic matter levels.

— *Preservation of soil structure.* All tillage destroys natural soil structure while no-tillage minimizes structural breakdown and increases organic matter and humus to begin the rebuilding process.

— *Preservation of earthworms and other soil fauna.* As with soil structure, tillage destroys humans' most valuable soil-borne ally, earthworms, while no-tillage encourages their multiplication.

— *Improved aeration.* Contrary to early predictions, the improvement in earthworm numbers, organic matter and soil structure usually result in improved soil aeration and porosity over time. Soils do not become progressively harder and more compact. Quite the reverse occurs, usually after 2–4 years of no-tillage.

— *Improved infiltration.* The same factors that aerate the soil result in improved infiltration into the soil. Plus residues reduce surface sealing by raindrop impact and slow down the velocity of runoff water.

— *Preventing soil erosion.* The sum of preserving soil structure, earthworms and organic matter, together with leaving the surface residues to protect the soil surface and increase infiltration, is to reduce wind and water soil erosion more than any other crop-production technique yet devised by humans.

— *Soil moisture conservation.* Every physical disturbance of the soil exposes it to drying, whereas no-tillage and surface residues greatly reduce drying. In addition, accumulation of soil organic matter greatly improves the water-holding capacity of soils.

— *Reduced irrigation requirements.* Improved water-holding capacity and reduced evaporation from soils lessen the need for irrigation, especially at early stages of growth when irrigation efficiency is at its lowest.

— *Moderating soil temperatures.* Under no-tillage soil temperatures in summer stay lower than under tillage. Winter temperatures are higher where snow retention by residue is a factor, but spring temperatures may rise more slowly.

— *Reduced germination of weeds.* The absence of physical soil disturbance under no-tillage reduces stimulation of new weed seed germination, but the in-row effect of this factor is highly dependent on the amount of disturbance caused by the no-tillage openers themselves.

— *Improved internal drainage.* Improved structure, organic matter, aeration and earthworm activity increase natural drainage within most soils.

— *Reduced pollution of waterways.* The decreased runoff of water from soil and the chemicals it transports reduces pollution of streams and rivers.

— *Improved trafficability.* Untilled soils are capable of withstanding vehicle and animal traffic with less compaction and structural damage than tilled soils.

— *Lower costs.* The total capital and/or operating costs of all machinery required to establish tillage crops are reduced by up to 50% when no-tillage substitutes for tillage.

— *Longer replacement intervals for machinery.* Because of reduced hours per hectare per year, tractors and advanced no-tillage drills are replaced less often and reduce capital costs over time. Some lighter no-tillage drills, however, may wear out more quickly than their tillage counterparts because of the greater stresses involved in operating them in untilled soils.

— *Reduced skills level.* While achieving successful no-tillage is a skilful task in itself, the total range of skills required is smaller than the many sequential tasks needed to complete successful tillage.

— *Natural mixing of soil potassium and phosphorus.* Earthworms mix large quantities of soil potassium and phosphorus in the root zone, which favours no-tillage because it sustains earthworm numbers and increases plant nutrient availability.

— *Less damage of new pastures.* The more stable soil structure of untilled soils allows quicker utilization of new pastures by stock with less plant disruption during early grazing than where tillage has been employed.

— *More recreation and management time.* The time otherwise devoted to tillage can be used to advantage for further management inputs (including the farming of more land) or for family and recreation.

— *Increased crop yields.* All of the above factors are capable of improving crop yields to levels well above those attained by tillage – but only if the no-tillage system and processes are fully practised without short cuts or deficiencies.

— *Future improvements expected.* Modern advanced no-tillage systems and equipment have removed earlier expectations of depressed crop yields in the short term to gain the longer-term benefits of no-tillage. Ongoing research and experience have developed systems that eliminate short-term depressed yields while at the same time raising the expectation and magnitudes of yield increases in the medium to longer term.

Disadvantages

— *Risk of crop failure.* Where inappropriate no-tillage tools and weed- or pest-control measures are used, there will be a greater risk of crop yield reductions or failure than for tillage. But where more sophisticated no-tillage tools and correct weed- and pest-control measures are used, the risks will be less than for tillage.

— *Larger tractors required.* Although the total energy input is significantly reduced by changing to no-tillage, most of that input is applied in one single operation, drilling, which may require a larger tractor or more animal power, or conversely a narrower drill.

— *New machinery required.* Because no-tillage is a relatively new technique, new and different equipment has to be purchased, leased or hired.

— *New pest and disease problems.* The absence of physical disturbance and retention of surface residues encourages some pests and diseases and changes the habitats of others. But such conditions also encourage their predators. To date, no pest or disease problems have proved to be insurmountable or untreat-able in long-term no-tillage systems.

— *Fields are not smoothed.* The absence of physical disturbance prevents soil movement by machines for smoothing and levelling purposes. This puts pressure on no-tillage drill designers to create machines that can cope with uneven soil surfaces. Some do this better than others.

— *Soil strength may vary across fields.* Tillage serves to create a consistently low soil strength across each field. Long-term no-tillage requires machines to be capable of adjusting to natural variations in soil strength that occur across every field. Since soil strength dictates the penetration forces required to be applied to each no-tillage opener, variable soil strength places particular demands on drill designs if consistent seeding depths and seed coverage are to be attained.

— *Fertilizers are more difficult to incorporate.* General incorporation of fertilizers is more difficult in the absence of physical burial by machines, but specific incorporation at the time of drilling is possible and desirable, using special designs of no-tillage openers.

— *Pesticides are more difficult to incorporate.* As with fertilizers, general incorporation of pesticides (especially those that require pre-plant soil incorporation) is not readily possible with no-tillage, requiring different pest-control strategies and formulations.

— *Altered root systems.* The root systems of no-tillage crops may occupy smaller volumes of soil than under tillage, but the total biomass and function of the roots are seldom different and anchorage may in fact be improved.

— *Altered availability of nitrogen.* There are three factors that affect nitrogen availability during early plant development under no-tillage.

— The decomposition of organic matter by soil microbes often temporarily 'locks up' nitrogen, making it less plant-available under no-tillage.

— No-tillage reduces mineralization of soil organic nitrogen that tillage otherwise releases.

— The development of bio-channels in the soil from earthworms and roots causes preferential flow of surface-applied nitrogenous fertilizers into the soil, which may bypass shallow, young crop roots.

— Each (or all) of these factors may create a nitrogen deficiency for seedlings, which encourages placing nitrogen with drilling. Fortunately some advanced no-tillage drills have separate nitrogen banding capabilities that overcome this problem.

— *Use of agricultural chemicals.* The reliance of no-tillage on herbicides for weed control is a cost and environmental negative but is offset by the reduction in surface runoff of other chemical pollutants (including surface-applied fertilizers) and the fact that most of the primary chemicals used in no-tillage are 'environmentally friendly'. Small-scale agriculture may require more hand weeding, but with greater ease than with tilled soils.

— *Shift in dominant weed species.* Chemical weed control tends to be selective towards weeds that are resistant to the range of available formulations, requiring more diligent use of crop rotations by farmers and commitment by the agricultural chemical industry to researching new formulations.

— *Restricted distribution of soil phosphorus.* Relatively immobile soil phosphorus tends to become distributed in a narrower band within the upper soil layers under no-tillage because of the absence of physical mixing. Improved earthworm populations help reduce this effect and also cycle nutrient sources situated below normal tillage levels.

— *New skills are required.* No-tillage is a more exacting farming method, requiring the learning and implementation of new skills, and these are not always compatible with existing tillage-related skills or attitudes.

— *Increased management and machine performance.* There is only one opportunity with each crop to 'get it right' under a no-tillage regime. Because no-tillage drilling is literally a once-over operation, there is less room for error compared with the sequential operations involved in tillage. This places emphasis on the tolerance of no-tillage drills to varying operator skill levels and their ability to function effectively in suboptimal conditions.

— *No-tillage drill selection is critical.*Few farmers can afford to own several different no-tillage drills awaiting the most suitable conditions before selecting which one to use. Fortunately more advanced no-tillage drills are capable of functioning consistently in a wider range of conditions than most tillage tools, making reliance on a single no-tillage drill for widely varying conditions both feasible and a practical reality.

— *Availability of expertise.* Until the many specific requirements of successful no-tillage are fully understood by 'experts', the quality of advice to practitioners from consultants will remain, at best, variable. Local, successful no-tillage farmers often become the best advisers.

— *Untidy field appearance.* Farmers who have become used to the appearance of neat, 'clean', tilled seedbeds often find the retention of surface residues ('trash') 'untidy'. But, as they come to appreciate the economic advantages of true no-tiilage, many such farmers gradually come to see residues as an important resource rather than 'trash' requiring disposal.

— *Elimination of 'recreational tillage'.* Some farmers find driving big tractors and tilling on a large scale to be recreational. Others regard it as a chore and health-damaging. Farmers in developing countries regard tillage as burdensome or impossible.

— The effects of no-tillage are developed as the soil and its physical and biological characteristics change. The result of these combined processes has been observed and documented in nearly every soil and climate worldwide, to the point of becoming common knowledge. It is in this transition stage that many who convert to no-tillage farming become disillusioned and sceptical that the benefits will in fact occur.

Benefits of No-tillage

Sustainable food and fibre production of any given field and region requires that the farming methods be economically competitive and environmentally friendly. To achieve this result requires adopting a farming technology that not only benefits production but provides an environmental benefit to the long-term maintenance of the soil and water resources upon which it is based. We must reduce pollution and use our resources in line with the earth's carrying capacity for sustainable production of food and fibre.

The responsibility of sustainable agriculture lies on the shoulders of farmers to maintain a delicate balance between the economic implications of farming practices and the environmental consequences of using the wrong practices. This responsibility entails producing food and fibre to meet the increasing population while maintaining the environment for a sustained high quality of life. The social value of an agricultural community is not just in production, but in producing in harmony with nature for improved soil, water and air quality and biological diversity.

Sustainable agriculture is a broad concept that requires interpretation at the regional and local level. The principles are captured in the definition reported by El-Swaify as: 'Sustainable agriculture involves the successful management of resources for agriculture to satisfy changing human needs, while maintaining or enhancing the quality of the environment and conserving natural resources.'

Conservation agriculture, especially no-tillage (direct seeding), has been proved to provide sustainable farming in many agricultural environments virtually around the world. The conditions and farming scales vary from humid to arid and vegetable plots to large prairie enterprises. All employ and adapt very similar principles but with a wide variety of machines, methods and economics.

The benefits of performing crop production with a no-tillage farming system are manyfold. Broad subjects discussed here only begin to provide the science and results learned over recent decades of exploring and developing this farming method. In addition to improved production and soil and water resource protection, many other benefits accrue. For example, it saves time and money, improves timing of planting and harvesting, increases the potential for double cropping, conserves soil water through decreased evaporation and increased infiltration, reduces fuel, labour and machinery requirements and enhances the global environment.

Principles of Conservation Agriculture

Conservation agriculture requires implementing three principles, or pillars. These are:

(i) minimum soil tillage disturbance;

(ii) diverse crop rotations and cover crops; and

(iii) continuous plant residue cover.

The main direct benefit of conservation agriculture and direct seeding is increased soil organic matter and its impact on the many processes that determine soil quality. The foundation underlying the three principles is their contribution and interactions with soil carbon, the primary determinant of long-term sustainable soil quality and crop production.

Conservation tillage includes the concepts of no-tillage, zero-tillage and direct seeding as the ultimate form of conservation agriculture. These terms are often used interchangeably to denote minimum soil disturbance. Reduced tillage methods, sometimes referred to as conservation tillage, such as strip tillage, ridge tillage and mulch tillage, disturb a small volume of soil and partially mix the residue with the soil and are intermediate in their soil quality effects. These terms define the tillage equipment and operation characteristics as they relate to the soil volume disturbed and the degree of soil–residue mixing. Intensive inversion tillage, such as that from mouldboard-ploughing, disc-harrowing and certain types of powered rotary tillage, is not a form of conservation tillage. No-tillage and direct seeding are the primary methods of conservation tillage to apply the three pillars of conservation agriculture for enhanced soil carbon and its associated environmental benefits.

True soil conservation is largely related to organic matter, i.e. carbon, management. By nothing more than properly managing the carbon in our agricultural ecosystems, we can have less erosion, less pollution, clean water, fresh air, healthy soil, natural fertility, higher productivity, carbon credits, beautiful landscapes and sustainability. Dynamic soil quality encompasses those properties that can change over relatively short time periods, such as soil organic matter, soil structure and macroporosity. These can readily be influenced by the actions of human use and management within the chosen agronomic practices. Soil organic matter is particularly dynamic, with inputs of plant materials and losses by decomposition.

Crop Production Benefits

Producing a crop and making an economic profit are universal goals of global farming. Production by applying no-tillage methods is no different in these goals, but there are definite benefits for the achievement. But these benefits only occur with fully successful no-tillage farming. There are certainly obstacles and risks in moving from traditional tillage farming, which has been the foundation technology for centuries.

Acceptable crop production requires an adequate plant stand, good nutrition and moisture with proper protection from weed, insect or disease competition. Achieving the plant stand in untilled, residue-covered soils is the first major obstacle, a particular challenge in modern mechanized agriculture, but certainly surmountable, as explained in the core of this text. Providing adequate nutrition and water for full crop potentials is readily achieved with the benefits of no-tillage, as discussed below.

Weed-control methods, by necessity, shift to dependence on chemicals, flame-weeding, mechanical crushing or hand picking for full no-tillage farming to stay within the goal of minimum soil disturbance. Chemical developments in recent decades have made great strides in their effectiveness, environmental friendliness and economic feasibility. Supplemental techniques of mowing, rolling and crushing without soil disturbance are showing significant promise to reduce weed presence and increase the benefit of cover crops and residues. Experience has shown that controlling insects and diseases has generally been less of a problem with no-tillage, even though there are often dire predictions about the potential impact of surface residues harbouring undesirables. As with weeds, crop health and pest problems are not likely to be avoided but may well shift to new varieties and species with the change in the field environment.

As a result of these developments and skilled applications, it has been repeatedly shown that crop production can be equalled and exceeded by no-tillage farming compared with traditional tillage methods. Because many soils have been tilled for many years, it is not uncommon to experience some yield reduction in the first few no-tillage years, largely because, as discussed later, it takes time for the soil to rebuild into a higher quality. This 'transition period reduction' can often be overcome or even averted with increased fertility, strategic fertilizer banding with drill openers and careful crop selection.

The full benefit of no-tillage comes in the reduced inputs. Most notable are the reduced inputs by minimizing labour and machine hours spent establishing and maintaining the crop. Reduced machine costs alone are significant, since all tillage equipment is dispensable. True no-tillage farming requires only an effective chemical sprayer, seeding–fertilizing drill and harvester.

With no seedbed preparation of the soil by tillage, seed drilling has become the major limitation to many efforts to successfully change to no-tillage farming. Modifying drills used in tillage farming has generally not been very successful, resulting in undesirable crop stands for optimum production. Many were not equipped to provide simultaneous fertilizer banding; thus it had to be provided by a supplemental minimum-tillage machine or, in the worst case, surface-applied, where it was very ineffective and stimulated weed growth.

Fortunately, drill development has progressed to now provide acceptable seeding in many cases, but many still do not fully meet all desirable attributes, especially in relation to the amount of soil disturbance they create. As a result of science and technique developments of recent years, no-tillage crop production now not only is feasible but has significant economic benefits. Combining and multiplying this result by the further benefits of soil and environmental qualities make no-tillage farming a highly desirable method of crop production. Further, many are now finding personal and social benefits from the reduced labour inputs, which remove much of the demanded time and drudgery often associated with traditional farm life. A common remark by successful no-tillage farmers is 'It has brought back the fun of farming.'

Increased Organic Matter

Understanding the role of soil organic matter and biodiversity in agricultural ecosystems has highlighted the value and importance of a range of processes that maintain and fulfil human needs. Soil organic matter is so valuable for its influence on soil organisms and properties that it can be referred to as 'black gold' because of its vital role in physical, chemical and biological properties and processes within the soil system.

The changes of these basic soil properties, called 'ecosystem services', are the processes by which the environment produces resources that sustain life and which we often take for granted. An ecosystem is a community of animals and plants interacting with one another within their physical environment. Ecosystems include physical, chemical and biological components such as soils, water and nutrients that support the biological organisms living within them, including people. Agricultural ecosystem services include production of food, fibre and biological fuels, provision of clean air and water, natural fertilization, nutrient cycling in soils and many other fundamental life support services. These services may be enhanced by increasing the amount of carbon stored in soils.

Conservation agriculture through its impact on soil carbon is the best way to enhance ecosystem services. Recent analyses have estimated national and global economic benefits from ecosystem services of soil formation, nitrogen fixation, organic matter decomposition, pest biocontrol, pollination and many others. Intensive agricultural management practices cause damage or loss of ecosystem services, by changing such processes as nutrient cycling, productivity and species diversity. Soil carbon plays a critical role in the harmony of our ecosystems providing these services.

Soil carbon is a principal factor in maintaining a balance between economic and environmental factors. Its importance can be represented by the central hub of a wagon wheel, a symbol of strength, unity and progress.

Based on soil carbon losses with intensive agriculture, reversing the decreasing soil carbon trend with less tillage intensity benefits a sustainable agriculture and the global population by gaining better control of the global carbon balance. The literature holds considerable evidence that intensive tillage decreases soil carbon and supports increased adoption of new and improved forms of no-tillage to preserve or increase storage of soil organic matter. The environmental and economic benefits of conservation agriculture and no-tillage demand their consideration in the development of improved soil carbon storage practices for sustainable production.

Increased Available Soil Water

Increased soil organic matter has a significant effect on soil water management because of increased infiltration and water-holding capacity. Enhanced soil water-holding capacity is a result of increased soil organic matter, which more readily absorbs water and releases it slowly over the season to minimize the impacts of short-term drought. Hudson showed that, for some soil textures, for each 1% weight increase in soil organic matter, the available water-holding capacity in the soil increased by 3.7% volume. Other factors being equal, soils containing more organic matter can retain more water from each rainfall event and make more of it available to plants. This factor and the increased infiltration with higher organic matter and the decreased evaporation with crop residues on the soil surface all contribute to improved water use efficiency.

Increased organic matter is known to increase soil infiltration and water-holding capacity, which significantly affect soil water management. Under these situations, crop residues slow runoff water and increase infiltration by earthworm channels, macropores and plant root holes. Water infiltration is two to ten times faster in soils with earthworms than in soils without earthworms.

Soil organic matter contributes to soil particle aggregation, which makes it easier for water to move through the soil and enables plants to use less energy to establish root systems. Intensive tillage breaks up soil structure and results in a dense soil, making it more difficult for plants to fully access the nutrients and water required for their growth and production. No-tillage and minimum-tillage farming allows the soil to restructure and accumulate organic matter for improved plant water and nutrient availability.

Reduced Soil Erosion

Crop residue management practices have included many agricultural practices to reduce soil erosion runoff and off-site sedimentation. Soils relatively high in C, particularly with crop residues on the soil surface, very effectively increase soil organic matter and reduce soil erosion loss. The primary role of soil organic matter to reduce soil erodibility is to stabilize the surface aggregates through reduced crust formation and surface sealing, resulting in less runoff. Reducing or eliminating runoff that carries sediment from fields to rivers and streams is a major enhancement of environmental quality. Under these situations, crop residues act as tiny dams that slow down water runoff from fields, allowing the water more time to soak into the soil.

Crop residues on the surface not only help hold soil particles in place but keep associated nutrients and pesticides on the field. The surface layer of organic matter minimizes herbicide runoff and, with conservation tillage, herbicide leaching can be reduced by as much as half.

Increased soil organic matter and crop residues on the surface will significantly reduce wind erosion. Depending on the amount of crop residues left on the soil surface, soil erosion can be reduced to near zero as compared with that from an unprotected, intensively tilled field. Wind or water soil erosion causes soil degradation and variability to the extent of a resulting crop yield decline.

Papendick *et al.* reported that the original topsoil on most hilltops had been removed by tillage erosion in the Palouse region of the Pacific Northwest of the USA. Mouldboard ploughs were identified as the primary cause, but all tillage implements will contribute to this problem. Soil translocation from mouldboard plough-based tillage can be greater than soil loss tolerance levels. Soil is not directly lost from the fields by tillage translocation; rather, it is moved away from the convex slopes and deposited on concave slope positions.

Lindstrom *et al.* showed that soil movement on a convex slope in southwestern Minnesota, USA, could result in a sustained soil loss level of approximately 30 t/ha/year from annual mouldboard-ploughing. Lobb *et al.* estimated soil loss in southwestern Ontario, Canada, from a shoulder position to be 54 t/ha/year from a tillage sequence of mouldboard-ploughing, tandem-discing and C-tine cultivating. In this case, tillage erosion, as estimated through resident caesium-137, accounted for at least 70% of the total soil loss. The net effect of soil translocation from the combined effects of tillage and water erosion is an increase in spatial variability of crop yield and a likely decline in soil carbon, related to lower soil productivity.

Enhanced Soil Quality

Soil quality is the fundamental foundation of environmental quality. Soil quality is largely governed by soil organic matter (SOM) content, which is dynamic and responds effectively to changes in soil management, tillage and plant production. Maintaining soil quality can reduce the problems of land degradation, decreasing soil fertility and rapidly declining production levels that occur in large parts of the world needing the basic principles of good farming practice.

Soil compaction in conservation tillage farming is significantly reduced by the reduction of traffic and increased SOM. Soane presented several mechanisms by which soil 'compactibility' can be affected by SOM:

1. Improved internal and external binding of soil aggregates.
2. Increased soil elasticity and rebounding capabilities.
3. Reduced bulk density due to mixing organic residues with the soil matrix.
4. Temporary or permanent existence of root networks.
5. Localized change of electrical charge of soil particle surfaces.
6. Change in soil internal friction.

While most soil compaction occurs during the first vehicle trip over the tilled field, reduced weight and horsepower requirements associated with no-tillage can also help minimize compaction. Additional field traffic required by intensive tillage compounds the problem by breaking down soil structure. Maintenance of SOM contributes to the formation and stabilization of soil structure. The combined physical and biological benefits of SOM can minimize the effect of traffic compaction and result in improved soil tilth.

While it is commonly known that tillage produces a well-fractured soil, sometimes requiring several tillage passes, it is a misconception that this is a well-aggregated, healthy soil. These soils never fare well when judged against modern knowledge of high 'soil quality'. A tilled soil is poorly structured, is void of many microorganisms and has poor water characteristics, just to name a few characteristics. As soils are farmed without tillage and supplied with residues, they naturally improve in overall quality, again support many microorganisms and become 'mellow' to the point of being easily penetrated by roots and earthworms. This transition takes several years to accomplish but invariably occurs given the opportunity.

Many traditional experienced farmers will often ask, 'How many years of no-tillage are possible before the soil becomes so compact as to require tillage?' No-tillage experience has shown exactly the opposite effect: once a no-tilled soil has regained its quality, it will continue to resist compaction and any subsequent tillage will cause undue damage. Most soils will continue to build organic matter and improve in quality criteria for years into the practice of no-tillage farming if the sequence is not broken by the thunderous effect of tillage.

Improved Nutrient Cycles

Improved soil tilth, structure and aggregate stability enhance the gas exchange and aeration required for nutrient cycling. Critical management of soil airflow, with improved soil tilth and structure, is required for optimum plant function. It is the combination of many factors that results in comprehensive environmental benefits from SOM management. The many attributes suggest new concepts on how we should manage the soil for long-term aggregate stability and sustainability.

Ion adsorption or exchange is one of the most significant nutrient cycling functions of soils. Cation exchange capacity (CEC) is the quantity of exchange sites that can absorb and release nutrient cations. SOM can increase this capacity of the soil from 20 to 70% over that of the clay minerals and metal oxides present. In fact, Crovetto showed that the contribution of organic matter to the cation exchange capacity exceeded that of the kaolinite clay mineral in the surface 5 cm of his soils. Robert showed that there was a strong linear relationship between organic carbon and the cation exchange capacity of his experimental soil. The capacity was increased fourfold with an organic carbon increase from 1 to 4%. The toxicity of other elements can be inhibited by SOM, which has the ability to adsorb soluble chemicals. Adsorption by clay minerals and SOM is an important means by which plant nutrients are retained in crop rooting zones.

Increased infiltration and concerns over the use of nitrogen in no-tillage agriculture require an understanding of the biological, chemical and physical factors controlling nitrogen losses and the relative impacts of contrasting crop production practices on nitrate leaching from agroecosystems. Domínguez *et al*. evaluated the leaching of water and nitrogen in plots with varying earthworm populations in a maize system. They found that the total flux of nitrogen in soil leachates was 2.5-fold greater in plots with increased earthworm populations than in those with lower populations. Their results are dependent on rainfall amounts, but do indicate that earthworms can increase the leaching of water and inorganic nitrogen to greater depths in the profile, potentially increasing nitrogen leaching from the system. Leaching losses were lower on the organically fertilized plots, attributed to higher immobilization potential.

Reduced Energy Requirements

Energy is required for all agricultural operations. Modern, intensive agriculture requires much more energy input than traditional farming methods since it relies on the use of fossil fuels for tillage, transportation, grain drying and the manufacture of fertilizers, pesticides and equipment used to apply agricultural inputs and for generating electricity used on farms. Reduced labour and machinery costs are economic considerations that are frequently given as additional reasons to use conservation tillage practices.

Practices that require lower energy inputs, such as no-tillage versus conventional tillage, generally result in lower inputs of fuel and a consequent decreases of CO_2-carbon emissions into the atmosphere per unit of land area under cultivation. Emissions of CO_2 from agriculture are generated from four primary sources: manufacture and use of machinery for cultivation, production and application of fertilizers and pesticides, the soil organic carbon that is oxidized following soil disturbance (which is largely dependent on tillage practices) and energy required for irrigation and grain drying.

A dynamic part of soil carbon cycling in conservation agriculture is directly related to the 'biological carbon' cycle, which is differentiated from the 'fossil carbon' cycle. Fossil carbon sequestration entails the capture and storage of fossil-fuel carbon prior to its release to the atmosphere. Biological carbon sequestration entails the capture of carbon from the atmosphere by plants. Fossil fuels (fossil carbon) are very old geologically, as much as 200 million years. Biofuels (bio-carbon) are very young geologically and can vary from 1 to 10 years in age and as a result can be effectively managed for improved carbon cycling. One example of biological carbon cycling is the agricultural production of biomass for fuel. The major strength of biofuels is the potential to reduce net CO_2 emissions to the atmosphere. Enhanced carbon management in conservation agriculture may make it possible to take CO_2 released from the fossil carbon cycle and transfer it to the biological carbon cycle to enhance food, fibre and biofuel production, for example, using natural gas fertilizer for plant production.

West and Marland conducted a carbon and energy analysis for agricultural inputs, resulting in estimates of net carbon flux for three crop types across three tillage intensities. The analysis included estimates of energy use and carbon emissions for primary fuels, electricity, fertilizers, lime, pesticides, irrigation, seed production and farm machinery. They estimated that net CO_2-carbon emissions for crop production with conservation, reduced and no-tillage practices were 72, 45 and 23 kg carbon/ha/year, respectively.

Total carbon emission values were used in conjunction with carbon sequestration estimates to model net carbon flux to the atmosphere over time. Based on US average crop inputs, no-tillage emitted less CO_2 from agricultural operations than did conventional tillage, with 137 and 168 kg of carbon/ha/year, respectively. The effect of changes in fossil-fuel use was the dominant factor 40 years after conversion to no-tillage.

This analysis of US data suggests that, on average, a change from conventional tillage to no-tillage will result in carbon sequestration in soil, plus a saving in CO_2 emissions from energy use in agriculture. While the enhanced carbon sequestration will continue for a finite time until a new equilibrium is reached, the reduction in net CO_2 flux to the atmosphere, caused by the reduced fossil-fuel use, can continue indefinitely, as long as the alternative practices are continued.

Lal recently provided a synthesis of energy use in farm operations and its conversion into carbon equivalents (CE). The principal advantage of expressing energy use in terms of carbon emission as kg CE lies in its direct relation to the rate of enrichment of atmospheric CO_2 concentration. The operations analysed were carbon-intensive agricultural practices that included tillage, spraying chemicals, seeding, harvesting, fertilizer nutrients, lime, pesticide manufacture and irrigation. The emissions for different tillage methods were 35.3, 7.9 and 5.8 kg CE/ha for conventional tillage, chisel tillage or minimum tillage and no-tillage methods of seedbed preparation, respectively.

Tillage and harvest operations account for the greatest proportion of fuel consumption within intensive agricultural systems.

Frye found fuel requirements using reduced tillage or no-tillage systems were 55 and 78%, respectively, of those used for conventional systems that included mouldboard-ploughing. On an area basis, savings of 23 kg/ha/year in energy carbon resulted from the conversion of conventional tillage to no-tillage. For the 186 million ha of cropland in the USA, this translates to a potential reduction in carbon emissions of 4.3 million metric tonnes carbon equivalent (MMTCE)/year.

These results further support the energy efficiencies and benefits of no-tillage. Conversion of ploughed tillage to no-tillage, using integrated nutrient management and pest management practices, and enhancing water use efficiency can save carbon emissions and at the same time increase the soil carbon pool. Thus, adopting conservation agriculture techniques is a holistic approach to management of soil and water resources. Conservation agriculture improves efficiency and enhances productivity per unit of carbon-based energy consumed and is a sustainable strategy.

Carbon Emissions and Sequestration

Tillage or soil preparation has been an integral part of traditional agricultural production. Tillage fragments the soil, triggers the release of soil nutrients for crop growth, kills weeds and modifies the circulation of water and air within the soil. Intensive tillage accelerates soil carbon loss and greenhouse gas emissions, which have an impact on environmental quality.

By minimizing soil tillage and its associated (CO_2) emissions, global increases of atmospheric carbon dioxide can be reduced while at the same time increasing soil carbon deposits (sequestration) and enhancing soil quality. The best soil management systems involve minimal soil disturbance and focus on residue management appropriate to the geographical location, given the economic and environmental considerations. Experiments and field trials are required for each region to develop proper knowledge and methods for optimum application of conservation agriculture.

Since CO_2 is the final decomposition product of SOM, intensive tillage, particularly the mouldboard plough, releases large amounts of CO_2 as a result of physical disruption and enhanced biological oxidation. With conservation tillage, crop residues are left more naturally on the surface to protect the soil and control the conversion of plant carbon to SOM and humus. Intensive tillage releases soil carbon to the atmosphere as CO_2, where it can combine with other gases to contribute to the greenhouse effect.

Soils store carbon for long periods of time as stable organic matter. Natural systems reach an equilibrium carbon level determined by climate, soil texture and vegetation.

When native soils are disturbed by agricultural tillage, fallow or residue burning, large amounts of carbon are oxidized and released as CO_2. Duxbury *et al.* estimated that agriculture has contributed 25% of the historical human-made emissions of CO_2 during the past two centuries. However, a significant portion of this carbon can be stored, or sequestered, by soils managed with no-tillage and other low-disturbance techniques. Increased plant production greater than that of native soil levels by the addition of fertilizers or irrigation can enhance carbon sequestration.

Carbon is a valuable environmental natural resource throughout the world's industrial applications of production and fossil energy consumption. Releasing carbon to the atmosphere by energy processes may be offset by capturing carbon with plant biomass and subsequently soil carbon sequestration in the form of organic matter. Energy consumers may at some time be required to compensate for their atmospheric carbon emissions by contracting with those who can sequester atmospheric carbon. Conservation agriculture may be able to provide this sequestration benefit and thus be compensated for its role in maintaining low net carbon emissions. While this 'carbon trading' mechanism is still in the discussion stage, it provides an important potential benefit.

Risk in No-tillage

The risks associated with no-tillage are those that result in reduced income to the farmer through impaired crop performance and/or increased costs. To be a sustainable technique, the failure rate for no-tillage must be no more, and preferably less, than that for tillage.

While early sceptics of the no-tillage concept forecast many and varied problems that would ultimately lead to the downfall of the practice, experience has shown that there are no insurmountable obstacles in most circumstances. The fact remains, however, that many farmers are still reluctant to attempt the new technique, fearing that it may increase their risks of crop failure or reduced yield.

The perception of risk is probably the single biggest factor governing the rate of adoption of no-tillage, and it is likely to remain so for a long time. Only education and personal experiences will finally put risk into perspective. Recent results convincingly show that no-tillage is not inherently more risky than conventional tillage, even in the short term. Indeed, it can reduce the risk factor during crop establishment if it is undertaken and managed correctly.

To plant and grow a crop with no-tillage, a farmer undertakes an economic risk that is affected by three functional risk categories: (i) biological; (ii) physical; and (iii) chemical. These risks are comparable between tillage and no-tillage systems because almost all of them are the everyday risks of cropping either way. Only their relative levels and remedies differ between the two techniques. The combined effects of the functional risks result in economic risks.

Biological Risks

Biological risks arise from pests, toxins, diseases, seed vigour, seedling vigour, nutrient stress and, ultimately, crop yield. The change to residue farming in general, which is the cornerstone of no-tillage, can have a marked effect on the incidence of diseases and pests, both positively and negatively. Seed placement and soil and residue disturbance by various drill or opener designs can influence all of these factors.

Pests

The change in earthworm and slug populations creates the most common pest problems in no-tillage. Slugs are particularly prone to proliferate in residue in high-humidity climates and must often be controlled by chemical means. Earthworms, on the other hand, can be either beneficial or damaging, depending on type. Earthworms generally provide positive effects that help aerate, drain and cycle nutrients. While tillage destroys earthworms, no-tilled soil nearly always has a significant and important increase in populations, and they are a great 'indicator' organism for other beneficial biota developments. Other damaging worms, such as wireworms, are generally not different regarding crop risks.

Slugs (Deroceras reticulatum) find shelter beneath the soil in many types of seed slots and feed on sown seeds and establishing seedlings. Clearly, slugs increase the biological risks of no-tillage. But they are relatively cheaply countered by the application of a suitable molluscicide.

Other pests can increase their damage risk because of increased surface residues or decreased physical destruction by tillage machines. But then so too do many of their predators.

An example of pest–drill interaction is that experienced with inverted T-shaped slots , which create subsurface soil-slot environments that are higher in soil humidity than either tilled soils or other no-tillage slots. Soil fauna that are sensitive to soil humidity, such as slugs and earthworms, tend to congregate in such slots. These may have both positive and negative effects for the sown crop.

Diseases

The most common soil disease that no-tillage appears to encourage is Rhizoctonia. Disturbance of the soil during tillage appears to partly destroy the fungal mycelia. Other fungal diseases are carried over in cereal residue and decaying organic matter in root channels, requiring diligent use of crop rotations or application of appropriate fungicides. On the other hand, the soil disease take-all (Gaeumannomyces graminis) appears to become more confined under no-tillage because of reduced soil movement. A concept called 'green bridge' was identified by Cook and Veseth , in which certain root bacteria from recent chemically killed plants can readily transfer to new seedlings if no-tillage seeding is undertaken within 14–21 days after the green crop begins dying. The specific pathogen has not yet been identified, but some delay after spraying and before no-tillage seeding appears to be an advantage where these bacteria exist, particularly in instances of continuous cereal cropping.

Toxins

The risks arising from toxins relate mainly to contact between seeds and decaying residue within the sown slot under persistently wet conditions . This risk, which is peculiar to no-tillage in cold wet soils, is eliminated by the use of no-tillage openers that effectively separate seed from the residues or the use of neutralizing agents sown with the seed.

The most common occurrence of residue effects has been experienced with double-disc drills seeding into wet, soft soils with surface residues. The residues tend to be folded and 'tucked' or 'hairpinned' into the seed slot with the seed dropped in the same location, which results in both the seed and residue experiencing decaying conditions and poor plant stands.

Some explanations for early no-tillage failures assumed that allelopathic exudates from dying plants may have killed newly sown seeds. But later detailed explanations for the causes of seedling emergence failures pointed to other (largely physical) factors and it has been hard to find any confirmed cases of allelopathy having played any role at all.

Nutrient stress

Without soil tillage to stir and mix applied fertilizer applications, careful attention must be paid to placing the fertilizer in untilled soils to optimize crop uptake and yield. Bands of fertilizer to the side and below the seed have proved to be very effective, sometimes utilizing one fertilizer band for each pair of seed rows. While it is important to place fertilizers far enough away from seeds and seedlings to avoid toxicity problems, it also appears that separation distances can (and indeed should) be much closer than those commonly accepted for tilled soils . Fertilizer banding has been found to be optimally accomplished by simultaneously seeding and fertilizing with a combination direct seed drill and fertilizer dispenser, and which is now common practice.

Again, the risk under no-tillage increases only if inappropriate equipment is used. On the other hand, there is voluminous evidence to show that, when fertilizers are placed correctly, no-tillage crop yields may be greater than those obtained from tilled soils . Thus, while the risk of nutrient stress under no-tillage may increase with inappropriate equipment, it may decrease compared with tillage if improved designs of no-tillage drills and planters are utilized.

Physiological Stress

It has been stated that untilled seedbeds are not as 'forgiving' as their tilled counterparts. This is often true because seedlings have to emerge through covering material that is physically more resistant than friable tilled soils. If the seeds are sown into mellow soils that have been no-tilled for several years or with scientifically designed furrow openers, such as inverted-T-shaped slots, the micro-environment of the slots will actually place less physiological stress on the seedlings than will a tilled soil. Thus physiological stress at the time of seedling emergence need not increase the biological risks. It may actually decrease the risk.

Seed quality

International seed testing authorities throughout the world test mainly for purity and optimally wetted germination as the main indicators of seed quality. But there are also agreed voluntary tests that describe other aspects of seed quality. One such test, the 'accelerated ageing' or 'vigour' test, examines a seed's ability to germinate after experiencing a period of stress (usually high or low temperature). It is possible for a given seed line to record a high-percentage germination but a low-percentage vigour. Therefore final germinations counts give no real indication of the vigour of a seed line although interim counts might be helpful in this respect.

There is an important interaction between seed vigour and drill opener designs, which can have important impacts on biological risk, and operators need to understand this interaction. No-tillage openers that create inverted-T-shaped slots produce about as favourable a micro-environment as it is possible to create for seeds, in either tilled or untilled soils. The main attribute is the availability of both vapour-phase and liquid-phase water. This ensures that even low-vigour seeds will germinate, almost regardless of the soil conditions.

In contrast, seeds sown into tilled soils or less favourable no-tillage slots that only provide liquid-phase water for germination of seeds are less likely to germinate. Farmers usually attribute such failures to a variety of reasons, but seldom test the vigour of the seed they had sown. When germination of low-vigour seeds does occur in tilled soils and open no-tillage slots, emergence of the seedlings is seldom restricted because of the friable nature of tilled soils and the open nature of vertical no-tillage slots. But the ensuing crop is likely to perform poorly.

Extensive field experience with inverted-T-shaped no-tillage slots, where even low-vigour seeds will often germinate under unfavourable conditions, have shown that the seedlings often did not have the vigour to emerge and were instead found twisted, weak and un-emerged beneath the soil surface. Observers at first attributed such twisting to fertilizer burn, but it is now known that fertilizer burn causes shrivelling and premature death of seedlings, not twisting. When vigour tests were carried out over a 3-year period on some 40 lines of seeds that had shown symptoms of subsurface seedling twisting in inverted-T no-tillage slots, all seed lines were found to be of low vigour (some as low as 18%).

The question is: What can be done about the problem? The responsibility rests with both the seed industry and individual no-tillage farmers. The seed industry needs to improve the quality of the seeds it offers for sale or at least be prepared to disclose information on seed vigour to farmers. Some companies already do this. No-tillage farmers, for their part, need to seek information from the seed industry about the vigour of particular seed lines and to be prepared to pay more for high-vigour lines. Those drill manufacturers that market advanced no-tillage seed drills need to advise purchasers that the weakest part of the system may now be seed quality, whereas previously it had been drill quality.

Physical Risks

Weather

Weather is likely to be the most variable and uncontrollable element in farming, and performing no-tillage won't change that. However, no-tillage does have the opportunity to significantly modify the impact by several means, some already mentioned or obvious. Increased available plant water is often the first noticeable effect, since residues and minimal soil disturbance reduce evaporation and increase infiltration.

Improved trafficability in wet soil is often a surprising no-tillage effect. With only one or two no-tillage crop years, the 'fabric' of the soil strengthens (mainly through improved soil structure) and animal or machine treading causes much less compaction with fewer surface depressions. It is common knowledge that no-tilled fields are accessible for seeding or spraying several days sooner following rainfall than tilled soils, with less damage by surface compaction. No-tilled soils are not more dense or compact than tilled soils; they just have more resistance to down pressures as a result of the increased organic matter and structure.

No-tillage also moderates excessive weather effects, such as extreme rainfalls and temperatures. With the surface residues protecting the surface against raindrop impact, runoff and erosion, rills and gullies don't form. Residues minimize the high wind profiles from having an impact on the soil surface and significantly reduce wind erosion. And very subtle dampening of soil temperature variations often prevents freezing of overwintering plants. No-tillage seeding into standing residues has allowed successful winter wheat crops in far more northerly climates in the northern hemisphere than previously possible, with increased yields compared with spring-seeded crops.

Young et al. showed how seasonal weather variations could affect the risk of altering the profitability of conservation tillage (which includes a component of no-tillage) compared with conventional tillage. They pointed out that the period 1986 to 1988 was particularly dry in the Palouse area of Washington State, which favoured the profitability of conservation tillage. The 1990/91 winter was particularly cold, which also favoured conservation tillage. At other times the weather did not favour either technique. In this manner the relative risks of changing profitability are clearly illustrated. Such risks cannot be predicted with any accuracy, but they can be minimized by selecting conservation tillage techniques and/or machines with the widest possible tolerance of changing weather patterns.

It is obvious that no-tillage machines cannot control the weather. But it has been repeatedly noted that when no-tillage is undertaken with appropriate residue manipulation and seeding machines designed with proper seeding slots, seeds and seedlings have considerably better protection from weather variations (e.g. too hot, cold, dry, windy or wet) than when that soil is either tilled or drilled with inappropriate no-tillage equipment. Thus, risks arising from inclement weather have the potential to be reduced under no-tillage if appropriate methods and equipment are used.

Machine function

Many of the physical risks arise from how well no-tillage machines perform their intended functions. The machine's designers must understand and incorporate the required capabilities to perform its intended functions in a wide variety of soil types, residues and weather conditions. These variations can change widely even within a single field or on a single day. There is much risk inserted into the farming system from a machine that operates at different levels of performance on different days in different parts of a field. A successful no-tillage drill must have a wide tolerance of changing, sometimes even hostile, conditions.

There are few more important physical functions than creating the correct micro-environment for the seeds within the soil. Different drill openers differ markedly in their abilities to do this and this affects the level of risk associated with different machines. To reduce machine-related risks, the openers of no-tillage drills must follow ground surface variations and move through significant surface residues without blockage. Seeding depth can only be maintained by careful tracking of the soil surface by the seed opener.

Maintaining surface residues is the main long-term benefit from no-tillage, especially for reducing erosion and temperature fluctuations and increasing soil fauna and infiltration. Residues are an equally important ingredient in short-term biological performance of seedling emergence and vigour. No-tillage does not offer the option to 'till out' last season's mistakes of vehicle ruts, animal paths, washed gullies, hardpans, etc. It is critically important to avoid creating field surfaces that are not mechanically manageable the following cropping season.

No-tillage seeding machines not only must physically handle residues consistently without blockage but must also have the ability to micro-manage those residues close to the slot and to utilize them for the benefit of the sown seeds and plants. Conversely, the inability of any opener to do these things significantly increases the risks from no-tillage, since the residues themselves are an important ingredient in creating a favourable habitat for seeds and seedlings. A positive utilization of crop residues in no-tillage is considerably different from tillage farming in that residues are seen as beneficial rather than a hindrance to machine performance. Since tilled soils, almost by definition, have minimal surface residues, they do not benefit in comparison with good utilization of residues by no-tillage openers, but they may compare well with no-tillage where residues are not utilized.

Similarly, the ability to uniformly track the untilled soil surface for uniform seeding by no-tillage drills will greatly determine the biological risks associated with poor seedling stands and vigour.

No-tillage drills encounter much higher forces and wear of components than their tillage counterparts. Since some of the critical functions, such as residue handling and slot formation, are often dependent on the mechanical wear remaining within narrow limits, maintenance of no-tillage machines is more important than for conventional drills. To put it another way, the absence of adequate maintenance on no-tillage drills may increase the risk of malfunction disproportionately.

None of the physical functions described above, however, has any relevance to risk unless its successful implementation has an identifiable biological function with regard to the sown seeds and emerging plants. Somewhat surprisingly, many of the early 'desirable functions' listed for no-tillage openers failed to define any biological objectives at all. Failure to recognize these biological-engineering linkages alone probably increased the level of risk of early no-tillage and accounted for much of the 'hit-and-miss' reputation the technique acquired in its early days.

The risk-assessment of the disc version of winged openers closely matches actual field surveys of users in New Zealand, which have consistently found a 90–95% success rate over several years and hundreds of thousands of hectares of field drilling. But the most commonly used opener throughout the world (vertical double disc) ranks poorly. This helps explain the many no-tillage failures associated with this opener.

Chemical Risks

Chemical risks have many of the same implications as physical risks. They are linked to the resultant biological risks that arise from them. Two stand-alone chemical risks are the effectiveness of weed control by herbicide application and the risk of toxicity or 'seed burn' from inappropriate placement of fertilizer in the seed slot relative to the seed placement.

Weed control

Weed control with herbicides must be as effective as that with mechanical means or the risk of impaired crop performance will increase. The principal variables determining herbicide effectiveness are as follows.

1 *Application of active ingredient.* The ability of operators to properly interpret the labels and literature supplied with various herbicides and pesticides has much to do with the success of applications. In addition, operators need to be able to recognize weed species and to be able to reliably calibrate their spraying machines. All of these operator choices are more risky than corresponding tillage operations. Nor are spraying mistakes as forgiving as tillage mistakes, which can often be 'repaired' the next day.

2. *Selection of appropriate chemical.* The selection of tillage tools can follow a trial-and-error routine where: (i) the non-performance of one implement becomes obvious within a short time; (ii) the consequences are seldom of great magnitude; and (iii) rectification using an alternative implement is accomplished quickly. Few, if any, of these flexibilities are available when choosing appropriate chemicals for a given weed or pest situation. Occasionally a mistaken choice can be rectified by the application of another chemical, but the options are fewer than with tillage and the risks are therefore greater.

3. *Weather.* Some chemicals require several hours without rain to be fully effective, while others are virtually 'rain-fast'. Since most chemicals involve a significant outlay of cash and, unlike tillage tools, are not

reusable, the risk from untimely rain and wind is greater than with tillage.

4. *Water quality.* Some foliage-applied herbicides, especially those that are inactivated upon contact with soil, such as glyphosate, have their efficacy altered by impurities in the mixing water. Of particular concern is water derived from storage dams or underground bores that is contaminated with particles or iron and carbonates. Some chemical effectiveness is quite variable with water acidity levels. Similarly, impurities on the leaves of target foliage, such as mud and dust from stock or vehicle traffic or recently applied lime, may inactivate some herbicides.
5. *Vigour of weeds.* The vigour of the target weeds at application time is important. Some herbicides (e.g. glyphosate) work best when sprayed on to healthy, actively growing plants. Others (e.g. paraquat) work best when the target plants are already stressed. Knowledge of these requirements is essential if effective weed control is to take place.
6. *Operator error.* During tillage, driving errors by an operator are seen immediately but they are seldom sufficiently serious to show up in the subsequent crop as an area of impaired yield. With once-over spraying, errors do not show up immediately. Paraquat is the most rapid to take effect but even then it is days before mistakes become visible. Most other herbicides take at least a week to show any visible effect, by which time the crop may have been sown, making remedial action virtually impossible without adversely affecting the sown crop.

Toxicity of Fertilizers

There are two risks from inappropriate fertilizer placement at sowing. If fertilizer isbroadcast on to the ground surface rather than placed in the soil at the time of drilling, there is a serious risk of impaired crop performance and yield as a result of limited plant availability . On the other hand, when fertilizer is sown with the seed there is a danger of the fertilizer damaging or 'burning' the seed under no-tillage unless the two are effectively separated in the soil. The latter risk increases with increased soil dryness. Separation is more difficult to achieve in no-tillage than in tilled soils, but it has been shown to be quite possible with the correct equipment without increased risk.

Economic Risk

All forms of risk during no-tillage are finally measured as economic risk. But economic risk should not be centred on cost savings alone. Indeed, focusing on cost savings may increase rather than decrease both real and imagined economic risks. This is for two reasons:

Where farmers already own tillage equipment, they see the acquisition of no-tillage equipment or even the use of contractors (custom drillers) – no matter how cheap – as duplication of an existing cost.

Purchasing inferior no-tillage equipment for cost savings may well result in lowered crop yields, even if only temporarily. Such a result may indeed be less cost-effective than either tillage or no-tillage undertaken with more expensive (and probably superior) equipment that maintains or even improves crop yields.

We shall examine both scenarios below.

The costs of tillage versus no-tillage

The costs of several alternatives for adopting no-tillage under a double cropping system (two crops per year, e.g. wheat followed by a winter forage crop for animal consumption) in New Zealand were analysed and compared with the costs of tillage. These were:

— Engaging a tillage contractor (custom driller) versus engaging a no-tillage contractor.

— Purchasing new tillage equipment versus purchasing new no-tillage equipment.

— Retaining ownership of used tillage equipment versus purchasing used no-tillage equipment.

— Retaining ownership of used tillage equipment versus purchasing new no-tillage equipment.

— Retaining ownership of used tillage equipment versus engaging a no-tillage contractor.

Fixed costs were included, such as interest on the investment, depreciation, insurance and housing, and expressed as a per-hour cost of annual machine use. Drills and planters are used for a shorter period each year to plant the same area under no-tillage than under a tillage regime. Thus the per-hour costs increase even though the per-hectare and per-year costs decrease. The analysis also assumed that a single large tractor and driver would be required for no-tillage compared with two or more smaller tractors and drivers for tillage.

For simplicity, the study assumed that the no-tillage drill being compared was of an advanced design, which ensured that crop yields would remain unchanged regardless of which option was chosen. Such an assumption is reasonable when applied to advanced no-tillage drills (which cost more anyway) but is unrealistic for inferior drills.

The cost analysis did not account for taxation issues, subsidies or other purchase incentives of any nature. These could otherwise be expected to favour no-tillage since many countries have incentives to encourage the practice because of its conservation value. Thus the results could be considered conservative in terms of the benefits recorded for no-tillage.

Farmers often see retention of their existing tillage equipment as 'insurance' while they gain the knowledge and skills necessary to master the new no-tillage technique to a stage where they can abandon tillage altogether. Other farmers claim that by going 'cold turkey' the learning process is achieved faster and more effectively.

Machine impacts on crop yields and economic risk

The effect of any one no-tillage drill design on crop yield and risk will be more important than its initial cost, when compared with either tillage or cheaper no-tillage alternatives. This belief has caused the research and development of improved no-tillage machines and systems as a means to reduce the risks associated with the practice, almost regardless of cost. The following analyses of machine capability versus expected crop yields and the resulting economics clarifies this belief.

The per-hectare charges that no-tillage contractors (custom drillers) make for their services are a good barometer of the relative costs associated with different no-tillage machines and systems. If we take New Zealand contractors as an example, we find that those with advanced (expensive) no-tillage drills in 2004 charged between US$72 and US$96/ha for their services, whereas those with lesser (cheaper) drills charged between US$36 and US$60/ha.

Differences between the ranges of charges are attributable mainly to differences in the initial costs of the two classes of machines and the different sizes of tractors needed to operate them. Differences within both ranges of costs reflect differences in the costs of competing options (such as tillage) together with differences in work rates and maintenance costs brought about by different field sizes, shapes, topographies and soil types (including abrasiveness). Taking the midpoint of each scale, the premium a farmer therefore paid in New Zealand in 2004 for access to a more advanced drill was about US$36/ha. Actual contractor charges in other countries will differ from these figures but the relativity between the costs associated with advanced machines and lesser machines is likely to be similar.

So a key question is: How much does an advanced no-tillage drill have to increase crop yields in order to justify the US$36/ha premium paid for the better technology under 2004 price conditions? Wheat sold in New Zealand in 2004 for approximately US$170/t. The average yield of spring-sown wheat in New Zealand in 2004 was 5.7 t/ha and the average autumn-sown wheat yield was 7.4 t/ha (N. Pyke, Foundation for Arable Research, 2004, personal communication). Gross returns for average spring- and autumn-sown wheat crops in 2004 were therefore US$969/ha and US$1258/ha, respectively.

To recover an additional US$36/ha in the costs of no-tillage drilling would require an increase in yield of 0.21 t/ha (or 210 kg/ha). This represented a 3.7% increase in yield of a spring-sown wheat crop or a 2.9% increase in an autumn-sown wheat crop. Such yield increases have been common. For example, the US Department of Agriculture obtained an average of 13% wheat yield increase in seven separate experiments over a 3-year period in Washington State by switching to a more advanced no-tillage drill compared with the best 'other' no-tillage drill that was then available. Similarly, the New South Wales Department of Agriculture (Australia) recorded an 11-year average of 27% yield advantage from soybean sown annually after oats using the same advanced no-tillage openers, compared with tillage. Commercial field experience over a 9-year period in New Zealand, the USA and Australia suggests that such research-plot measurements have been a realistic reflection of field expectations. Wheat and other crop yields approaching twice the national averages have become common from no-tillage practised at its most advanced level.

4

Resource-conserving Technologies in Rice-based Systems

In 2000, about 850 million people were suffering from hunger, 815 million of which in developing countries. From a global perspective, hunger is a problem not of food production but of accessibility, since to date world food production has kept pace with demand. Although recent predictions point to reduced population growth, a considerable increase in overall production is still required—a major challenge, given the already stretched land and water resources. Future yield increases may be limited compared to past trends, but by 2030, food production must double to keep pace with demand.

The production of renewable resources is increasingly important, due both to the growing awareness of sustainability and to rising oil prices. Countries producing a food surplus could in future focus more and more on the production of profitable renewable raw material for industrial products and energy rather than simply on the production of food.

There are also regional differences to be taken into account. Sub-Saharan Africa remains a hunger hot spot with a stagnating trend in per capita food production, while other regions are seeing a steady increase. Countries with a large population like China and India, and which in the past achieved food self-sufficiency, now face the challenge of maintaining and increasing high yield levels in a scenario of increasing climatic variability.

Conserving Natural Resources for Agricultural Production

Over the last few decades the growth in agricultural production has come mainly from yield increase and to a lesser extent from area expansion. Now the agricultural land available per capita is expected to decline while revolutionary technologies for significantly higher production potential do not seem to be in sight. Furthermore, in high intensity agricultural production areas, yield increase seems to have reached a ceiling despite higher input use; in some cases, yields even decline, for example in the grain-producing areas of Punjab in India.

Water is one of the most precious natural resources for agricultural production and agriculture accounts for 70 percent of water use. It is predicted that by 2025 water consumption will exceed "blue water" availability if current trends continue. In the Indian state of Punjab, characterized by intensive irrigated agriculture, the groundwater table is falling at a rate of 0.7 m per year. However, the decline of freshwater resources is due not only to increased consumption, but to careless management. Agriculture contributes to the problem by wasting water and by sealing and compacting the soils so that excess water cannot infiltrate and recharge the aquifer—one of the causes of the growing number of flood catastrophes. In regions where water is already the limiting factor for agricultural production, this wasteful practice threatens the sustainability of agriculture. Rising temperatures and evapotranspiration rates combined with more erratic rainfall further aggravate the water problems in rainfed agriculture.

Soil affects not only production, but also the management of other natural resources, such as water. Soil structure is strongly correlated with the organic matter content and the soil life. Organic matter stabilizes soil aggregates, provides feed to soil life and acts as a sponge for soil water. With intensive tillage-based agriculture, the organic matter of soil is steadily decreasing, leading first to a decline in productivity, followed by the visible signs of degradation and finally desertification. The lack of yield response to high doses of fertilizer in the Indo-Gangetic Plains can be attributed to poor soil health resulting from over-exploitation. In the Indian states of Uttaranchal and Haryana, the organic carbon content in soils reaches minimum values of less than 0.1 percent. While soil degradation is more pronounced in tropical regions, it is also a phenomenon in moderate climatic zones; indeed, the world map of degraded soils indicates that nearly all agricultural lands show some level of soil degradation.

Bed Planting

Bed planting refers to a cropping system where the crop is grown on beds and the irrigation water is applied in furrows between the beds. This is common practice for row crops, but not for small grain crops such as wheat and rice. The technique offers a number of advantages, such as improved fertilizer efficiency, better weed control and reduced seed rate. It also saves irrigation water compared to a flat inundated field: the evaporation surface is reduced and water application and distribution efficiency are increased. In addition, the rooting environment is changed and the aeration of the bed zone is better than in flat planting. Reported water savings (compared to flat surfaces) reach 26 percent for wheat and 42 percent for transplanted rice, and yield increases 6.4 percent for wheat and 6.2 percent for rice.

Resource-conserving technologies (RCT) have been developed in order to:

— reduce the use of and damage to natural resources through agricultural production; and

— increase the efficiency of resource utilization.

Most of these technologies target the two most crucial natural resources: water and soil, but some also affect the efficiency of other production resources and inputs (e.g. labour, farm power and fertilizer). Some of the more popular RCTs, particularly in irrigated or rice-based cropping systems, are described below.

Laser Levelling

For surface irrigated areas it is essential to have a properly levelled surface with the appropriate inclination for the irrigation method adopted. Traditional farmers' methods for levelling by eyesight are not sufficiently accurate, resulting in extended irrigation times, unnecessary water consumption and inefficient water use. The use of laser-guided equipment for the levelling of surface-irrigated fields has become economically feasible and accessible—through hiring services—even to lower-income farmers. Laser levelling reduces the unevenness of the field to about ±2 cm, resulting in better water application and distribution efficiency, improved water productivity, increased fertilizer efficiency and reduced weed pressure. Savings of up to 50 percent in wheat and 68 percent in rice have been reported.

Direct Seeding

Direct seeding of rice—compared with transplanting -may be considered an RCT:

— It saves labour and fuel.

— Seeding into dry soil saves water as there is no puddling.

— The total growing period from seed to seed is reduced by about 10 days.

— Yields and water efficiency of the subsequent rotation crops are increased.

On the other hand, weed management is more difficult in dry direct-seeded rice than in puddled and transplanted rice.

Reduced Tillage, Zero Tillage

Intensive soil tillage is the main cause of reduced soil organic matter and hence of soil degradation. Tillage accelerates the mineralization of organic matter and destroys the habitat of the soil life. On the contrary, when soil tillage is reduced or eliminated, soil life returns and the mineralization of soil organic matter slows down, resulting in better soil structure. Under zero tillage the mineralization of soil organic matter can be reduced to levels inferior to the input, converting the soil into a carbon sink. Zero tillage also results in water saving and improved water-use efficiency: since the soil is not exposed through tillage, the unproductive evaporation of water is reduced while water infiltration is facilitated. The potential water saving through zero tillage varies according to the cropping system and the climatic conditions. On average, water savings of 15 to 20 percent can be expected. Used in isolation, however, zero tillage can lead to problems with weed control, compaction or surface crusting, depending on the soil type.

Mulching and Green Manure

The supply of organic matter to the soil through mulching and green manure is important for maintaining and enhancing soil fertility. Mulching material can come from crop residues or green manure crops; it provides feed for the soil life and mineral nutrients for the plants. If legume crops are used as green manure they can supply up to 200 kg/ha of nitrogen to the soil; in the case of rice, this can result in mineral fertilizer savings of 50 to 75 percent. The spreading of mulch on the soil surface reduces evaporation, saves water, protects from wind and water erosion, and suppresses weed growth.

Controlled Traffic Farming

Controlled traffic farming restricts any traffic in the field to the same tracks. While these tracks become heavily compacted, the rooting zone does not at all, resulting in better soil structure and higher yields. The area lost in the traffic zones is easily compensated for by better growth of plants adjacent to the tracks so that overall yields are usually higher than in conventional systems with random traffic. Controlled traffic farming is the ideal complement to zero tillage or to bed planting systems, but it also provides advantages in conventional agriculture through time and fuel savings, since resistance to soil tillage in the compaction-free rooting zones is significantly lower and traction is more efficient when tyres work on compacted tracks. In the latter case, either GPS (global positioning system) guidance or visible bed and furrow systems must be used to limit the tillage operation to the rooting zones and to not disturb the tracks.

Synergies Between Resource-conserving Technologies

Resource-conserving technologies provide scope for synergy, for example, between bed planting and controlled traffic or mulching and zero tillage. Used in isolation, any of these technologies may face specific problems (e.g. surface crusting or weeds in direct seeding rice) or have limitations (e.g. zero tillage under irrigated conditions).

The combination of resource-conserving technologies working in synergy is commonly referred to as "conservation agriculture" (CA).

Conservation Agriculture

Conservation agriculture (CA) is a concept for resource-saving agricultural crop production that strives to achieve acceptable profits with high and sustained production levels while conserving the environment. CA is based on enhancing natural biological processes above and below the ground. Interventions such as mechanical soil tillage are reduced to an absolute minimum, and external inputs—e.g. agrochemicals and nutrients of mineral or organic origin—are applied at an optimum level taking care to not interfere with or disrupt the biological processes. CA is characterized by three interlinked principles:

— minimum mechanical soil disturbance throughout the entire crop rotation;
— permanent organic soil cover; and
— diversified crop rotations in the case of annual crops or plant associations in the case of perennial crops.

During the last decade, CA has been gaining in popularity throughout the world and is now applied on about 95 million ha. Together with other organizations and stakeholders, FAO has been promoting and introducing CA in several countries in Latin America, Africa and Asia. CA adapts to different climatic conditions from the equatorial tropics to the vicinity of the polar circle and to different crops and cropping systems, including vegetables, root crops and paddy rice.

Zero Tillage

When the soil is not tilled, the soil structure changes. A system of continuous macropores is established, facilitating water infiltration and soil aeration as well as root penetration into deeper zones. Soil organic matter content increases, with higher values near the surface, gradually decreasing with depth. Soil macro- and microfauna and flora is re-established resulting in better soil fertility.

Soil Cover

The permanent soil cover through crops, mulch or green manure cover crops complements the effects of zero tillage by supplying substrate for soil organic matter buildup and for the soil life which is facilitated by not disturbing the soil. By protecting the soil surface, the mulch reduces evaporation, avoids crusting and suppresses weed growth. Problems experienced in direct seeding or zero tillage (applied in isolation) are thus reduced. It should also be noted that the application of zero tillage and direct seeding facilitates the management of residues which in conventional systems are often considered a problem. cropping system (i.e. conservation agriculture). The investment in laser levelling lasts much longer than in conventional systems, since under CA no further soil tillage (which could upset the levelling of the field) is applied.

Crop Rotation

In addition to the phytosanitary and weed management benefits, crop rotation serves to open different soil horizons with different rooting types. While conventional agriculture "cultivates the land", using science and technology to dominate nature, conservation agriculture tries to "least interfere" with natural processes. Similar thoughts have been developed over the past 50 years in the Far East by Masanobu Fukuoka and applied in rice-based farming.

Permanent Beds

In systems where surface irrigation is applied, bed planting results in water saving. Under CA, the beds are converted into permanent beds and soil tillage is limited to a periodic cleaning and reshaping of the furrows. The same permanent bed system is applicable under conservation agriculture also for crop rotations, which include crops grown on beds (e.g. for drainage purposes). The furrow distances and bed width must be harmonized for all crops in the rotation and for all mechanized traffic operations. In this way a permanent bed system leads also to controlled traffic, another RCT.

Direct Seeding

Direct seeding is another complement to CA. Although transplanting of crops, including paddy rice, is possible under zero tillage, direct seeding is preferable for the reasons mentioned above. Direct seeding results in less soil movement than transplanting, which often involves some sort of strip tillage. CA facilitates direct seeding by reducing a number of problems encountered when direct seeding is applied in isolation (e.g. surface crusting or weed control).

Laser Levelling

The benefits of laser levelling in conservation agriculture are as for conventional agriculture under surface irrigation conditions. To begin with, significant soil movement is required, and so laser levelling is considered an initial investment before converting to a permanent zero tillage

Effects of CA

Under CA the levels of soil erosion are inferior to the build-up of new soil. The soil under CA "grows" at an average rate of 1 mm per year due to the accumulation of soil organic matter. This growth continues until a new point of saturation is reached in the soil which takes 30 to 50 years. The organic matter levels rise by 0.1–0.2 percent per year due to the residues left on the soil surface, the remaining root biomass and the reduced mineralization. Within a crop rotation, different root systems structure different soil horizons and improve the efficiency of the soil nutrient use. In general the soil structure becomes more stable.

Soils under conservation agriculture also improve water efficiency. The increased amount of continuous vertical macropores facilitate the infiltration of rainwater into the ground and help recharge the aquifer. The increased soil organic matter levels improve the level of water accessibility to plants: 1 percent of organic matter in the soil profile can store water at a rate of 150 m/ha. The permanent soil cover and the avoidance of mechanical soil tillage reduce the unproductive evaporation of water, water-use efficiency is increased and a crop's water requirements can be reduced by about 30 percent under both irrigation and rainfed conditions. In addition to the quantitative benefits, reduced leaching of soil nutrients and farm chemicals together with reduced soil erosion lead to a significant improvement in the water quality in watersheds where CA is applied.

CA can reduce the overall requirement for farm power and energy for field production by up to 60 percent compared to conventional farming. This is due to the fact that the most power-intensive operations, such as tiilage, are eliminated and equipment investment, particularly the number and size of tractors, is significantly reduced. CA produces a decline in the use of agrochemicals due to enhanced natural control processes: natural control of pests and diseases improves over time and experience in weed management through crop rotations also facilitates

this long-term decline in agrochemical use. The same is true for mineral fertilizer: less fertilizer is lost through leaching and erosion and the different rooting systems recycle more soil nutrients from a larger soil volume, resulting in improved overall efficiency of fertilizer use in the long term with a significant reduction in the fertilizer requirements to maintain the production and soil nutrient levels over the crop rotation.

Climate and Climate Change

Recent decades have seen an increase in the frequency and strength of harsh climatic events, including very high precipitations as well as extended drought periods and extreme temperatures. Agricultural production systems are highly vulnerable to these changes.

Conservation agriculture can assist in the adaptation to climate change, by improving the resilience of agricultural cropping systems and making them less vulnerable to abnormal climatic situations. Better soil structure and higher water infiltration rates reduce the danger of flooding and erosion following high intensity rainstorms. Increased soil organic matter levels improve the water-holding capacity and hence the ability to cope with extended drought periods. Yield variations under CA in extreme years (dry or wet) are less pronounced than under conventional agriculture.

But CA also helps mitigate the effects of climate change, at least with regard to the emission of greenhouse gases. With the increasing soil organic matter, soils under CA can retain carbon from carbon dioxide and store it safely for long periods of time. This carbon sequestration continues for 25 to 50 years before reaching a new plateau of saturation.

The consumption of fossil fuel for agricultural production is significantly reduced under CA and burning of crop residues is completely eliminated, which also contributes to a reduction in greenhouse gas release. Soils under zero tillage—depending on the type of management—might also emit less nitrous oxide. With paddy rice in particular, the change to zero tillage systems combined with adequate water management can positively influence the release of other greenhouse gases, such as methane and nitrous oxides.

Cropping System

Irrigated paddy rice has for a long time been considered a stable and sustainable cropping system, although it is far from being conservation agriculture (the puddling results in the destruction of the soil structure). However, irrigated rice is increasingly subject to pressures:

— The high fuel costs of puddling and the reduced availability of labour mean that there is pressure to change from transplanted to direct-seeded rice.

— The water consumption of traditionally puddled rice is too high in many regions—alternatives must be found and rice growing is already restricted in some areas: cultivation of summer rice, grown prior to the monsoon season, is not allowed in parts of northern India; in Karakalpakstan, adjacent to the Aral Sea in Uzbekistan, rice cultivation is restricted because of the scarce water resources and the high evaporation losses; in China, the paddy rice areas around the city of Beijing have been replaced by other crops due to the alarming fall in the groundwater table.

— The release of greenhouse gases such as methane is high in traditionally flooded rice.

Rice cultivation has therefore been adapted to conservation agriculture in several countries. Rice can be cultivated without puddling or permanent flooding by adopting resource-conserving technologies. FAO has been working on rice-based CA systems in China and the Democratic People's Republic of Korea, while in the Indo-Gangetic Plains the Rice-Wheat Consortium has been successfully introducing RCTs into rice-based cropping systems. Neither puddling nor zero tillage in rice result in higher yields of the non-rice crops in the crop rotations. The reported water saving through RCTs is usually higher in paddy rice than in other rotation crops. Cropping systems involving residue retention and zero tillage perform better in terms of profitability, yields and resource conservation, while conventional systems and zero tillage systems without residue retention are inferior. In addition to the resource-conserving effects, the cropping systems involving permanent zero tillage, so-called "double zero tillage" and residue retention result in significantly increased water infiltration rates.

The experiences and results obtained in CA in other cropping systems can be confirmed for rice-based cropping systems. This includes options for mitigating climate change by sequestering carbon in the soil and reducing the emission of other greenhouse gases. The Rice-Wheat Consortium has developed technologies which allow the application of conservation agriculture in rice-based cropping systems: laser levelling, permanent bed planting and the retention of residues (including rice straw). In rice-wheat systems, the introduction of sesbania as a cover crop to bridge the gap between the wheat harvest and rice seeding is well accepted by the farming community. It helps with weed control and adds additional nitrogen and organic matter to the system. Direct seeding equipment has been developed and introduced to the market to seed different crops into residues and under zero tillage either on flat fields or raised beds. The latest model of the "Turbo Happy Seeder" can even cope with seeding into fresh rice straw.

Resource-conserving technologies applied in isolation have advantages and disadvantages; they are not universally applicable as the problems can sometimes outweigh the benefits. However, by combining different resource-conserving technologies, synergies can be created to eliminate the disadvantages of single technologies and accumulate the benefits.

Different RCTs are successfully applied under the concept of conservation agriculture in different cropping systems around the world, allowing stable agricultural production without the known negative environmental impact. The Rice-Wheat Consortium of the Indo-Gangetic plains has been instrumental in adapting the concept of conservation agriculture to rice-based cropping systems, resulting in higher yields, greater profitability, enhanced soil fertility and better water-use efficiency—it represents a possible route towards sustainable agricultural production in rice-based systems. High water consumption is a particular concern. In regions where cropping mainly depends on groundwater for irrigation purposes and where the groundwater tables are falling dramatically, such as in the Punjab of India, water-saving technologies might not be sufficient to guarantee sustainability of the cropping systems. The combination of different RCTs –

such as mulching, direct seeding and double zero tillage—results not only in water saving but also in increased infiltration rates (and the staff recharge of the aquifer during the monsoon season).

Combined resource-conserving technologies applied in conservation agriculture produce benefits for the farming sector, the environment and the general public, and it is therefore important to promote and adopt them. FAO and regional partners, such as the Rice-Wheat Consortium, can play an important role in this process.

ENERGY GENERATION FROM RICE RESIDUES

Reliable and adequate supplies of energy are a fundamental necessity for economic growth and development of any community or country.

Technologies for Energy Production from Rice Residues

Properties of importance in energy generation

Rice straw and rice husk are lignocellulosic materials with a low bulk density and a relatively high silica content. The main chemical components on a dry basis are cellulose, hemicelluloses, lignin and ash. Lignin is found in the middle lamella and adjacent primary cell walls of the residue tissue, and as such it encapsulates the cellulose and hemicellulose fractions found primarily in the secondary cell walls. Rice straw and husk have relatively high proportions of silica-rich ash.

Steps in the energy conversion chain

The chain of operations involved in obtaining energy from rice residues comprises the following five steps:

1. collection, handling and delivery to a pre-treatment site;
2. pre-treatment to prepare the biomass for subsequent conversion;
3. conversion of the residue to fuel;
4. refining of the fuel; and
5. conversion of the fuel into usable energy.

The conversion technology applied dictates the pathway taken to implement each of the five steps in the chain. However, irrespective of which pathway *Pre-treat*ment is taken, the principal technological challenge that must be faced is to cost-effectively achieve an optimal net positive energy balance for the chain while minimizing the negative effects on the environment and fostering the economic and social development of local communities. As energy—mostly provided by fossil fuels—is required to carry out operations in the chain, an appropriate technology package is required involving minimal energy input (in order to maintain a positive energy balance) and resulting in minimal emission of greenhouse gases and other atmospheric pollutants.

Collection and handling of rice residues

Residues are collected, handled and transported to the site where subsequent operations are carried out. For each kilogram of rice grain, 0.25 kg of husks and 1.25 kg of straw residues are generated. Husks are found in high concentrations at the mill where dehusking takes place and can therefore be readily collected and used for energy generation to power various mill operations. In comparison, straw is usually strewn over wide areas in harvested fields and may require substantial inputs of energy—often from fossil fuels—for baling, temporary storage, drying and transportation to the pre-treatment or conversion site.

Pre-treatment turns residues into a form that facilitates handling, improves efficiency and limits the adverse environmental impacts of conversion processes. The main pre-treatment steps are described below:

— *Drying* may be done naturally—sun-drying—or by artificial means in an appropriately designed drier. Size reduction processes (e.g. chopping and maceration) increase the surface area available for, and hence the rate of, biochemical conversion reactions. Given the high resistance of lignin to biodegra-dation, delignification is required to expose the cellulose and hemicellulose fractions to biochemical conversion reactions. One method is to steep in an alkali (e.g. sodium hydroxide) and then wash with water.

— *Densification* is applied to residues intended for direct combustion and gasification. It improves combustion characteristics, increasing the conversion efficiency and reducing environmental pollutants. It facilitates the handling, transportation and storage of residues. Rice straw can be densified by baling during collection, or by size reduction followed by briquetting; rice husks can be densified by briquetting, after (or without) prior size reduction.

Conversion

Several technologies exist for converting rice residues into usable energy, including: thermochemical conversion through direct combustion, gasification and fast pyrolysis; and biochemical conversion through anaerobic digestion and fermentation.

1. *Thermochemical conversion processes*

   — *Direct combustion.* Heat energy is produced during combustion of rice straw or rice husks in stoves, furnaces and kilns. These devices must be properly designed to ensure proper operation, optimal conversion efficiency and heat efficiency. The feeding system must be designed to meter residue at an optimal rate into the combustion chamber. Densification into pellets improves the handling and combustion characteristics of residues. Rice residues have a high ash content and provision must be made to evacuate ash as fast as it forms in order not to impede the airflow into the combustion area. The design and maintenance plan must take into account the fact that the heat transfer efficiency will be decreased by ash deposits on the combustion chamber walls and that, due to its high silica content, ash is very abrasive and corrosive of metals.

   — *Gasification.* Gasification is the thermal conversion of rice straw and rice husk into producer gas (a combustible mixture of carbon monoxide, hydrogen, methane, nitrogen and carbon dioxide). Compared with direct combustion, gasification gives greater flexibility in the use of the fuel and higher conversion efficiency. A residue of ash is formed in the process, and so the design of the combustion chamber is as for direct combustion.

— *Fast pyrolysis (bio-oil production).* Bio-oil is a type of liquid fuel formed through the condensation of gases generated by the decomposition of rice residues by fast pyrolysis. It is a complex mixture of hydrocarbons, phenolic materials and water—the latter acts as a diluent to maintain homogeneity and optimal viscosity. Fast pyrolysis is an advanced process with carefully controlled parameters ensuring high heat transfer rates to the feedstock; a crucial pre-treatment step is size reduction of the residue to very finely divided pieces. Fast pyrolysis for conversion of biomass to organic liquids in high yields became a technical reality 25 years ago.

2. *Biochemical conversion processes:*

— *Fermentation (ethanol production).* Biochemical conversion processes require two main pre-processing operations—size reduction and delignification. Following these operations, the production of ethanol from rice residues involves two main steps: enzymatic hydrolysis of cellulose and hemicellulose into simple hexose and pentose sugars, and fermentation of simple sugars into ethanol. In recent years there has been a marked acceleration in the development of economically feasible technologies for lignocellulosic ethanol production. These technologies are now commercialized and numerous commercial plants are being developed in the United States, Canada, Brazil, Europe and Japan.

— *Anaerobic digestion.* Anaerobic digestion of biomass in semi-continuous digesters is one of the bioenergy generation technologies that have been applied successfully in small-scale operations in rural areas of developing countries. Anaerobic digestion generates biogas, a gaseous mixture of methane, carbon dioxide, hydrogen with small quantities of carbon monoxide, oxygen and hydrogen sulphide. A carbon-to-nitrogen ratio of around 30 is the threshold above which it is considered that the quantity of nitrogen available in the digestion feedstock may be insufficient to support the metabolic processes of the microbiological populations involved. With carbon-to-nitrogen ratios that exceed this value by far, rice residues are unsuitable as anaerobic digestion feedstock; however, they may be used if the nitrogen content is raised by mixing with appropriate nitrogen-rich materials (e.g. animal manure, night soil and ammonia). Ammonia also has a delignification effect on ligno-cellulosic materials, and ammoniation may therefore be carried out to achieve both delignification and nitrogen-content adjustment of rice residues. The fibrous nature of rice residues may pose a problem, as residues tend to float and form a hard scum on the surface of digester contents; in this case, special measures must be taken—maceration of the slurry before feeding into the digester and intermittent or continuous mechanical mixing during digestion.

Energy utilization

Fuels obtained from rice residues can provide energy in various forms: heat, motive power and electricity.

Heat: Heat energy produced during direct combustion of rice residues can be used for domestic cooking and heating, as well as for providing process heat in small- and medium-scale industrial operations (e.g. rice parboiling and drying, tobacco curing, bricks and ceramics production, pottery etc.). It is also possible to generate heat by burning liquid fuels (bio-oil and ethanol) and gaseous fuels (producer gas and biogas) derived from residues). Heat generated through direct combustion of residues, or through burning liquid and gaseous derivative fuels, can be used to generate boiler steam to provide heat for use in domestic or industrial operations.

Mechanical power: Gaseous fuels (biogas and producer gas) and liquid fuels (bio-oil and ethanol) obtained from rice residues can be used in internal combustion engines to provide shaft power for pumps, crop processing machinery (e.g. dryer fans and grain threshers), refrigeration systems and other devices in rural areas.

Liquid fuels offer the advantage of easy storage and transportation, and are therefore more suitable than gaseous fuels for powering vehicles and mobile equipment. Unlike ethanol, bio-oil does not mix easily with conventional fossil fuels and has to substitute, rather than be blended with, gasoline or diesel fuel in engines.

Producer gas and biogas are cleaned by bubbling through scrubbers and filters before being used in engines. Both gases contain high proportions of carbon dioxide which is incombustible. Biogas also contains hydrogen sulphide—a corrosive substance that increases engine wear and burns to form sulphur dioxide, itself corrosive and a contributor to acid rain. During gasification, tars and particulates (ash and char) get entrained in the resulting producer gas. They cause damage to engines and lead to environmental pollution.

Electricity: There are three ways of obtaining electrical power usingrice residues as energy feedstock:

— Gaseous or liquid fuel derivates can be used to run an internal combustion engine providing shaft power to an electricity generator.

— Gaseous fuel derivatives can be used in a gas turbine running an electricity generator.

— Steam generated in a boiler can be used to operate a turbine that in turn runs an electricity generator.

Electricity generated using any of these three methods can be used in decentralized or grid-connected systems for domestic or industrial applications.

Combined heat and power (CHP): Rice residues can be used in CHP (cogeneration) systems to simultaneously generate heat and electricity by using either flue gases from electricity-generating gas turbines or waste steam from electricity-generating steam turbines to provide heat for industrial or domestic use. Cogeneration systems are in wide use in rice processing plants, where rice husk from milling is used in boiler furnaces for raising steam which is in turn used for generating electricity to power milling operations. The steam which exits the steam turbine is used to provide process heat for drying and parboiling operations.

Utilization of by-products

The commercial utilization of by-products formed in the bioenergy chain increases the economic feasibility and commercial viability of systems for generating energy from rice residues.

Direct combustion and gasification

A substantial quantity of ash is generated as a by-product during direct combustion and gasification. The ash is rich in silica, is highly porous and has a very high surface area. It is a good heat insulator and is therefore suitable for incorporation in refractory bricks and for ensuring uniform cooling and solidification of steel. It can also be used as a fertilizer, oil absorbent, anti-caking agent, filter medium, filler in rubber components, additive in cement and as a source of silica for industrial operations such as glass making.

Anaerobic digestion

The effluent from semi-continuous anaerobic digesters can be used in soil fertility improvement and conditioning, as well as in animal feed and fishmeal.

Cellulosic ethanol production

The main by-product from ethanol production is lignin from delignification. With a heating value of 26.63 MJ/kg, lignin is a bona fide fuel material that can be used to provide industrial process heat. Lignin can also be used as raw material for producing a variety of high-value chemicals such as phenols.

Bio-oil production

Non-condensable gases and the char residue formed during the fast pyrolysis of rice residues can be used as stand-alone fuels to provide heat in various industrial operations.

Overcoming Constraints in the Use of Rice Residues

The various constraints to the adoption and use of technologies for converting rice residues and other agricultural biomass into energy in rural areas in developing countries are outlined below and the action to be taken by various stakeholders is highlighted. Case examples from Asia and Africa are then presented.

Policies and institutional framework

In many countries there are no policies providing an environment conducive to the successful development and operation of energy technologies based on biomass in general and rice residues in particular.

Governments must formulate policies and programmes for the bio-energy sector, as part of wider strategic plans for economic development and improvement of living standards in rural areas. Such policies should take into account the multifaceted nature of residue energy production and utilization, with consideration for the technical, economic and socials aspects.

A framework of relevant institutions is required to support residue energy systems. The framework should involve all government ministries and public institutions dealing with the relevant sectors (i.e. agriculture, energy, rural development, environment, planning, industrial development, public works, infrastructure etc.). It should also have links with relevant private sector partner institutions, including financial institutions, advisory services, NGOs, community organizations and utility companies. The policy and institutional framework must develop and enforce rules and regulations for: guiding production, marketing and utilization of energy; providing criteria for quality grades and standards; minimizing negative environmental and health impacts; and applying various pricing instruments and incentives such as subsidies and taxes. International organizations have an important role in providing financial and technical assistance in the development of institutional frameworks, strategies, policies and programmes.

Infrastructure

Many countries lack the infrastructure base required to support the development and utilization of bioenergy technologies.

Governments (local and national) are responsible for providing the infrastructure and infrastructure services that are public goods, and they must also provide the necessary environment and incentives for private sector investment in other infrastructures (residue storage structures, fuel storage structures, pipelines etc.).

Where feasible, local communities should be involved in the construction and maintenance of infrastructure, for example, laying pipe networks to distribute biogas from community digesters to homes. Assistance from international agencies is required to undertake the analysis of infrastructure needs, identify suitable infrastructure and operational models, mobilize required financial resources, develop and implement guidelines and standards at international level, and provide technical and financial assistance in infrastructure development programmes.

Awareness, data and information There is a lack of awareness of the various aspects of energy generation from rice residues among feedstock producers, technical support personnel, researchers, private sector individuals, policy-makers and energy users. They are not familiar with the technical options for energy generation or the advantages compared to conventional fuels, and they are unaware of the opportunities for job creation and income generation, as well as the social and environmental implications. When people are aware, they are nevertheless without reliable and accurate information concerning critical elementssuch as potential resources, energy demand, energy utilization and related socio-economic factors.

Governments and private sector stakeholders should carry out awareness-raising campaigns and pilot demonstration projects; they should take appropriate action to generate the data necessary for technology selection and design, policy formulation and development planning. In addition, they should disseminate information dealing with residue utilization, using the available media (printed bulletins and newsletters, radio, television, Internet etc.) and targeting specific interest groups or the public at large.

Through normative (and other) studies, international organizations need to generate relevant data and information; they should also carry out activities within programmes and projects to facilitate access to information and strengthen information exchange networks.

Capital

A very important factor inhibiting the widespread adoption of residue transformation and utilization technologies in rural areas is the lack of capital to cover investment and operations costs.

Measures are required to improve the access of rural dwellers, community groups, potential business people and micro-enterprises to credit and finance schemes that are suitable for local conditions. Governments need to set up an institutional framework and incentives (e.g. low interest or subsidized loans) in order to encourage investment by the private sector. International organizations should provide assistance in developing appropriate institutional frameworks and microcredit schemes.

Skills

The adoption and application of technologies required is hindered by the low level of technical and business management skills and lack of organizational capacities in rural areas. The people who plan the development of bioenergy systems and those that provide technical support for operating these systems often lack the necessary technical and managerial skills.

Training programmes should be organized by governments and the private sector to impart the required skills to policy-makers and planners, and to develop a critical mass of technical personnel familiar with the technologies. Training programmes are required to target relevant rural dwellers, community groups and micro-enterprises and to cover the basic technical aspects, small business management and marketing. International organizations can provide financial and technical assistance in projects with components related to training and demonstration.

Technology

Affordable, easy-to-operate, economically viable, socially acceptable and environmentally friendly technologies adapted to local conditions are required for the production and utilization of energy from rice residues. Basic and applied research is needed to develop new technologies or adapt existing ones to local conditions. Research is required with regard to the technical aspects (e.g. improving conversion efficiency), environmental impacts and socio-economic implications of technology packages, and an appropriate institutional framework must be put in place to support the research, development and demonstration of technologies. The private sector is central to technology development: it carries out its own investigations, provides grants for research and collaborates with research and teaching institutions in their activities. International agencies can provide technical and financial assistance to facilitate technology and knowledge transfer and to strengthen national capacity in assessing, developing and adapting technologies.

Case example 1: Improved rice husk furnace for rice parboiling in Bangladesh

Biomass is by far the dominant energy source in Bangladesh, accounting for approximately 67 percent of the country's total energy consumption. A survey of rice mills revealed that an average 187 kg of husks are produced per tonne of paddy and 70 percent of husks are consumed by the mills themselves for energy generation. Traditionally, mills use poorly designed furnaces to power boilers, leading to poor thermal efficiency and high pollution. Carbon monoxide levels in exhaust gases are double the threshold established by the Ministry of Environment and Forestry.

Collaboration between the Bangladesh Rice Research Institute and the Natural Resources Institute (United Kingdom) led to the design of an improved system featuring: better thermal efficiency, improved fuel conversion efficiency, shorter parboiling time, and a level of carbon monoxide in the flue gas well within legally permitted limits. Studies carried out by a national research organization and technical assistance provided by a development agency were instrumental in the development of an economically viable and environmentally sustainable rice residue conversion technology.

Case example 2: Setting policy in Viet Nam

In 2003, Viet Nam passed a decree (Prime Minister's Decree No.102/2003/ND-CP) to guide the rational exploitation of energy resources in meeting the growing energy demands of the national economy while protecting the environment and achieving sustainable socio-economic development. The decree outlined the legal instruments (e.g. tax preferences) for the importation of energy-saving technologies, and directed ministries and other public sector agencies to allocate funding for scientific, technological and environment research targeting the efficient use of energy. The decree also required these agencies to disseminate relevant information using the mass media and to carry out awareness-raising activities.

Case example 3: Incentives for cellulosic alcohol in China

China imports close to 43 percent of its petroleum requirements and in 2005 petroleum imports increased to about 865 million barrels. In order to reduce dependence on imports, the Government is taking measures to promote the use of ethanol as a petroleum substitute. Given the food security implications of obtaining ethanol from corn, cellulosic ethanol was identified as a more viable substitute for petroleum than corn alcohol.

In 2005, cellulosic ethanol was selected as one of the key environmental protection and energy development technologies to receive priority support under the national strategic high-technology research and development programme. In order to attract foreign technology and capital, the Government recently announced plans to invest US$ 5 billion over the next 10 years in ethanol capacity expansion with a focus on cellulosic ethanol (Yang and Lu, 2007—based on).

Case example 4: Technical assistance by FAO in Egypt

Through imaginative agronomic programmes and technical assistance to growers, Egypt's rice production systems have become some of the highest-yielding in the world. While high yields have secured the availability of rice grain for feeding the population and earning foreign exchange, large quantities of residue are generated, constituting a major disposal problem. Almost 2 million tonnes of rice straw are burned on-farm resulting in greenhouse gas emissions, aerial pollution and the loss of potential revenue that could be generated from processes using straw as raw material.

In December 2006, FAO started a 20-month pilot technical cooperation project (TCP) in the Nile Delta; one of the key specific objectives is to analyse the technical and economic feasibility of using rice straw as feedstock for energy generation, and findings will be used to design strategies within a proposal for a follow-up project. This is an example of how an international organization can successfully provide technical and financial assistance to support studies, information generation and the design of future interventions in a national programme.

Rice residues have immense potential as an energy source in rural areas of developing countries. Technologies applied in the conversion and utilization chain should result in favourable carbon and energy balances and in minimal particulate and greenhouse gas emissions. To overcome the constraints to the adoption and use of these technologies in rural areas, action is required by the public sector, the private sector, local communities, development agencies and international organizations. An appropriate policy, institutional and regulatory framework is needed to create an environment that is conducive to the development and operation of these technologies. Awareness needs to be raised about energy generation from rice residues, while access to capital and relevant information needs to be facilitated. The technical and managerial capacity of various stakeholders needs to be strengthened, and support has to be provided for research, development and demonstration of low-cost technologies appropriate for the particular locality.

Weed Management in European Rice Fields

Rice is cultivated in the European Union on submerged land in the coastal plains, deltas and river basins, covering a total area of about 400 000 ha, all in the Mediterranean countries. The average crop yields are between 4.80 and 7.25 tonnes/ha, depending on the environmental conditions and water availability. Milled rice consumption ranges from about 6.0–7.0 kg/caput (17 kg/caput in Portugal) in the Mediterranean areas, where the crop is traditionally grown, to 3.5–5.3 kg/caput in non-rice-producing countries. In Europe's rice-producing countries, most of the rice consumed is from *japonica* varieties, while in northern countries mostly *indica* varieties are consumed.

The ecological conditions of rice cultivation vary with climates ranging from temperate to subtropical. At higher latitudes (Italy, northern Spain, France) rainfall is concentrated during the first stages of the crop (April–June) and during the harvesting period. Average temperatures in these areas range from 10°–12°C during crop germination to 20°–25°C at flowering time.

Rice is mostly grown on fine-textured, poorly drained soils, where the pH is between 4 and 8 and organic matter between 0.5 and 10 percent. In coastal areas, soils are frequently saline or very saline. Most of the irrigation water is obtained from rivers (Po in Italy, Ebro and Guadalquivir in Spain, Tejo in Portugal, Axios in Greece etc.) and lakes. In European areas, rice is mostly cultivated under permanent flooding, adopting mainly the "flow-through" system, with short periods during which the soil is dried to favour rice rooting (in the early stages) or weed control treatments. Water is kept at a level ranging from 5 cm (in the early stages) to 12 cm (from tillering to flowering).

Rice is mostly cultivated as a monocrop in the same field for many years. Seedbeds are usually prepared by ploughing in autumn or spring at a depth of 20 cm, incorporating residues of the previous crop into the soil. Rice is mainly directly seeded, broadcasting seeds in flooded fields with fertilizer spreaders or by airplane (in Spain).

Fertilizer rates are typically:

— N: 80-120 kg/ha (50% in pre-planting and 50% in post-planting, using urea or other ammoniac fertilizers)

— P: 60-80 kg/ha (pre-seeding)

— K: 100-150 kg/ha (pre-seeding)

Weed Scenarios

Weeds are considered the worst noxious organisms affecting rice production in Europe. It is estimated that without weed control, at a yield level of 7 to 8 tonnes/ha, yield loss can be as high as 90 percent. Weed problems are mainly related to competition with rice plants for light and nutrients, resulting in yield and quality reduction and an increase in the cost of harvesting and drying. Herbicides account for more than 80 percent of the total consumption of pesticides used in crop protection, with a total spending of about 110 million euros per year.

The important changes that occurred in most of the Western European countries during the 1960s in rice management—such as the change from transplanting to direct seeding, the expansion of mechanization and the introduction of chemical weed control—brought about a significant modification in the composition of the weed flora in rice fields.

The rice-field ecosystem is notably complex and the numerous weed species (both C3 and C4) present are characterized by particular morphophysiological traits. C plants are mainly found in dry-seeded fields, while C species tend to dominate in submerged rice crops. Depending on the specific ecological conditions and anthropic pressure, some species may appear while others can disappear over time. The major weeds growing in European rice fields are aquatic; they may be grouped on the basis of the practices adopted to control them, as follows:

— *Echinochloa* species: *E. crus-galli, E. crus-pavonis, E. oryzoides, E. erecta* and *E. phyllopogon.* Major weeds in rice-cropping systems worldwide, in both water and dry-seeded rice (Holm *et al.,* 1977; Ferrero *et al.,* 2002), they have high variability in morphological and competition-related traits (e.g. plant size, tillering ability, seed dimensions and germination behaviour), which makes field identification of different species difficult and uncertain.

— *Heteranthera* species: *H. reniformis, H. rotundifolia* and *H. limosa.* Exotic plants first reported in Italy in 1962, in some areas they have become widespread in recent years and can compete severely with rice from its early stages.

— Alisma sedges and the sedges group (cyperaceae weeds) *Alisma plantago-aquatica, A. lanceolatum, Cyperus difformis, Bolboschoenus maritimus, Schoenoplectus mucronatus* and *Butomus umbellatus.* They are grouped together, as they are normally subject to common control programmes and are often sensitive to the same herbicides.

— Weeds in drill-seeded fields from two different floristic groups related to the different ecological conditions present on dry and flooded soil. Dry soil: *Echinochloa* spp, *Panicum dichotomiflorum, Bidens* spp, *Digitaria sanguinalis, Polygonum* spp, *Chenopodium album* and *Amaranthus retroflexus.* Flooded soil: weed species reported in fields flooded since rice planting. In these cultural conditions specific programmes of control are required for both groups of weeds.

— Weedy rice biotypes of cultivated rice *(Oryza sativa* L.). Diffused in much of the world, it is estimated that weedy rice infestations cover between 40 and 75 percent of European rice fields and in Italy, France and Spain weedy rice infestations have been reported on 60 to 75 percent of the rice cultivated area.

At the seedling stage, weedy rice plants are difficult to distinguish from the crop, while after tillering the identification of the weed is possible thanks to many distinct morphological differences from the rice varieties: more numerous, longer and more slender tillers; leaves which are often hispid on both surfaces; tall plants; pigmentation of several plant parts; and easy seed dispersal after formation in the panicle. Weedy rice grains frequently have a red pigmented pericarp and the term "red rice" is commonly adopted in international literature to identify these spontaneous plants. This is not very appropriate, however, as red coat grains are also present in some cultivated varieties and absent in various weedy forms.

When the seeds break off and onto the soil prior to crop harvesting, the weeds disseminate and feed the soil seedbank. Following the shift from rice transplanting to direct seeding, red rice spread; the last 15 years have seen the problem worsen in Europe with the cultivation of weak, semi-dwarf *indica*-type rice varieties. The current spread is mainly related to the planting of commercial rice seeds containing grains of the weed.

Wee Management

All crop management practices may determine the competitive ability of rice and the weeds infesting rice fields. The shift in the early 1950s from transplanting to direct seeding and the abandoning of manual weeding resulted in greater infestations of weeds, including *Echinochloa* spp, *Alisma* spp, sedges and weedy rice plants. Weed management was complicated further following the introduction of short-stature rice varieties and the practice of shallow water in the fields during the early stages which created an ecological environment more favourable to their growth.

A sustainable programme of weed management must be based on a combination of cultural and chemical means: neither chemicals nor cultural practices alone can give satisfactory weed control.

Cultural management

The main cultural operations that can have a significant

direct or indirect impact on weed management are described below.

Soil tillage

Land preparation is an important component of rice weed control programmes: it helps the establishment and growth of rice while suppressing or delaying the development of weeds.

Tillage carried out in the autumn or winter increases soil aeration, favours straw decomposition and reduces algae infestations the following year. Established perennial weeds can then be partially devitalized if the soil dries out before field flooding for seeding the crop. Where perennial weeds are present, equipment fitted with rotary organs of tillage must not be used.

With soil tillage, fertilizers can be incorporated at a depth of 5 to 10 cm, reducing their availability to weeds which can germinate at the soil surface, and limiting nitrification and losses of nitrogen. With minimum tillage, weed seeds remain in the upper layers of the soil; they are spread uniformly and control is therefore more effective—both in pre-seeding (adopting the stale seed bed technique) and in early post-emergence.

Land levelling

Precision land levelling, obtained with laser-directed equipment, has made an important contribution to weedy rice management in European rice production. Level or regularly sloping fields enable appropriate water management, which limits weed growth and guarantees uniform emergence of weeds, which in turn makes herbicides more effective. With good soil levelling, the basin is larger and there are fewer ditches and levees from which weeds can spread into the fields.

Water management

Water management is central to weed control in water seeded rice. The water depth in rice fields has changed quite remarkably over the years, depending on the different growth features of the new varieties introduced in time (different vegetative vigour during the early stages, tall or short size of the plants, tillering degree etc.) and on the specific requirements of the herbicides applied.

In levelled fields, water is maintained at a depth of 5 to 7 cm until tillering and then at 12 to 15 cm until a few days before ripening. Rice fields are commonly drained 2 or 3 times during the crop cycle, in order to:

— hasten rooting of the rice seedlings;

— oxygenate the soil to avoid risks of unfavourable fermentation in the first few days after seed germination (7-15 days after seeding);

— destroy the algal scum;

— apply herbicides requiring drained soil conditions; or

— spread nitrogen fertilizers.

Most herbicides introduced into the market in recent years are characterized by foliar absorption and the plant surface must be well exposed to the herbicide spray. Dry conditions stimulate weed germination; therefore draining should be short to avoid the creation of different-aged weeds which are difficult to control with herbicides.

Rotation

Rotating rice with dry crops is an effective means of managing weeds that cannot be successfully controlled in rice. Where there are high infestations of weedy rice, the rotation of rice with non-flooded crops is the best solution to get a significant reduction of the seed-bank of this weed.

System of rice planting

On a small percentage of the cultivation area, rice is planted in dry soil and only flooded from the beginning of tillering until ripening; the rice fields are then infested by two different floristic groups related to the different ecological conditions of dry and flooded soil. While planting in dry soil reduces or delays the growth of those weed species requiring an aquatic environment (e.g. *Heteranthera* spp), it increases the development of non-aquatic weeds (e.g. *Panicum dichotomiflorum*, *Digitaria sanguinalis* and *Polygonum* spp). Weeds can be partially removed from dry fields with one or two passes of the tine harrow.

Varieties

The cultivation of *indica*-type and early-maturing varieties has significantly increased over recent years in European countries. Most *indica*-type varieties are short, have low growth and are only moderately competitive with weeds—all features which contribute to the spread of weedy rice infestations.

Early varieties usually have a cycle of 130–145 days (about 20–30 days shorter than regular varieties); they are popular because they escape the negative effects of the low temperatures in April and August when the delicate phases of emergence and flowering occur. High-yielding varieties planted in mid-May and which begin to flower before August are favoured. Where short-cycle varieties are selected, weedy rice control is carried out before rice planting.

Herbicide management

Herbicides are a fundamental element in sustainable weed management programmes. Numerous herbicides are available to control major rice weeds. In recent years much effort has gone into developing herbicide programmes that maximize the use of commercial products while reducing the number of treatments.

Herbicide strategies are established mainly on the basis of the composition of the infestations. The key factors to be considered when deciding the weed control programme are the seeding conditions and the presence of weedy rice; the latter may influence the organization of the cultural practices or the choice of the herbicides.

Water seeding: infestations of Echinochloa spp, Heteranthera spp, Alisma spp, sedges and others

When weedy rice is absent and infestations are characterized by the presence of most common weeds (e.g. *Echinochloa* spp, *Heteranthera* spp and ciperaceae), two or three treatments are commonly required: one in pre-emergence (mainly against *Heteranthera* spp) and one or two 10–40 days after crop emergence.

The first treatment is normally: oxadiazon (300-380 g / ha) (to control *Heteranthera* species); combined with a graminicide (against *Echinochloa* plants, e.g. thio-bencarb or molinate); and sometimes with an ALS inhibitor applied at one-half or one-third the normal rate (against sedges and other weeds). They are applied 3-4 days before planting and soil flooding, or 5-6 days before planting on flooded soil.

Second and sometimes third treatments are frequently necessary to control late emergences of *Echinochloa* spp, sedges, alismataceae and other species.

Water seeding infestations with weedy rice, Echinochloa spp, Heteranthera spp, Alisma spp, sedges and others

Weed control programmes are principally aimed at weedy rice control and are carried out prior to seeding:

— with an antigerminative herbicide (e.g. flufenacet or pretilachlor) applied about 1 month before rice planting; or

— by mechanical means (harrows or a tractor fitted with cage wheels) or with systemic graminicides (e.g. dalapon, cycloxydim, clethodim or glyfosate) to destroy weedy rice seedlings grown after stale seedbed application.

In both cases, a second treatment is usually required to control *Heteranthera* spp in rice pre-planting, and at least a third treatment with a mixture of specific herbicides, to control *Echinochloa* spp, sedges and other weeds, 25-40 days after rice planting.

When pre-planting treatments against weedy rice are not carried out or are unsatisfactory, there is often an intervention at rice flowering time. This can be manual when there are only a few weedy rice plants per ha, but with high infestations, the weed is devitalized by applying systemic herbicides (glyphosate) with wiping bars, provided that the weed plants are taller than those of the crop. The Clearfield® technology is particularly promising: it is based on the planting of a rice variety tolerant to imazamox, an imidazolinone herbicide with a wide spectrum of activity that also includes weedy rice plants.

Drill seeding infestations with Echinochloa spp, Panicum dichotomiflorum, Digitaria spp, Polygonun spp, Alisma spp, sedges and others

When rice is seeded in dry soil, two or three treatments are generally required:

— First, in rice pre-emergence: pendimethalin (1 000-1 300 g /ha) or clomazone (200-230 g /ha) to control *Echinochloa* spp and other weed grasses; when *Heteranthera* spp is present, oxadiazon is usually added.

— Second, 10-30 days after rice emergence: propanil in combination with an ALS inhibitor to control sedges, *B. umbellatus, A. plantago-aquatica* and *Echinochloa* spp plants which escaped the pre-emergence treatment.

— Third, if necessary, just before flooding: against weeds which escaped previous treatments or emerged late, this intervention is sometimes performed after field flooding, applying the same products used in flooded rice fields.

Technological advances in the equipment mean that herbicides are sprayed with low water volume and at low pressure; this improves the efficiency of the products and limits the risk of environmental pollution. Furthermore, thanks to the increased width of the boom sprayers, fewer passes are made in the basins by tractors equipped with toothed wheels, thus diminishing both the cost of spraying and the frequency of late weed germinations which can occur along the wheel tracks.

When using herbicides, particular attention must be paid to water movement and depth in the rice field. Most ALS inhibitors require static water for a few days in order to avoid chemical removal and allow uniform soil absorption. Foliar herbicides (propanil, MCPA etc.) should be applied on drained fields for maximum exposure of the weed foliage to the spray. The improper use of herbicides can result in the appearance of resistant species, cause environmental pollution and risk disrupting the precarious balance of natural pest enemies.

The principal resistant weeds belong to *S. mucronatus, A. plantago-aquatica, C. difformis* and *Echinochloa* spp. Studies of *C. difformis* and *S. mucronatus* have shown that there is a generalized cross resistance among several sulfonylureas (azimsulfuron, bensulfuron-methyl, cinosulfuron, imazamox and byspiribac-sodium). Some resistant populations are also insensitive to triazolopyrimidine herbicide (metosulam) at three times the recommended field dose.

The main techniques currently adopted by European rice growers to tackle herbicide resistance are the rotation of herbicides and the application of mixtures of herbicides with different modes of action. For the successful control of ALS-resistant *Alisma* and sedges, farmers are increasingly using hormonic herbicides, such as MCPA which was widely used before the introduction of ALS inhibitors. Crop rotation—probably the best preventive and curative method for dealing with herbicide resistance—is unlikely to be adopted by farmers, for technical (soil suitability for other crops), economic and organizational reasons.

Environmental contamination from herbicide use is an important issue requiring attention, with particular regard to the choice of active ingredients having low solubility in water, low volatilization and low persistence.

Weed management is a major concern for rice growers, as unsuccessful weed control can result in a severe reduction in yield and quality.

The dramatic technological advances of recent decades have influenced rice management: the development of mechanization; the introduction and diffusion of chemical weed control; the change from transplanting to direct seeding; and the introduction of late, dwarf and less competitive rice varieties. All these changes determined important modifications in the composition of the weed flora in rice fields. Weeds such as *Echinochloa* spp, *Alisma* spp, cyperaceae species and weedy rice—previously well controlled by hand-picking or limited in their growth by transplanting—became increasingly competitive. The introduction of rice seed from other countries favoured the diffusion of exotic weeds such as *Heteranthera* spp. The numerous herbicides now available, and which are suited to every floristic situation, help limit yield losses but often do not prevent weed spread and pressure. The improper use of herbicides may lead to the development of resistance in some weeds or to environmental pollution. The main issues in rice weed management can be addressed by integrated strategies based on an appropriate combination of herbicides with good agronomic practices.

Rice Genetic Potential and its Application in Rice Breeding for Stress Tolerance

Using the vast genetic potential of the rice collection and modern methods of creation and evaluation of the breeding material, the All-Russia Rice Research Institute (ARRRI) has released a number of varieties resistant to environmental stress factors, well adapted to local conditions and widely used. ARRRI also provides a valuable source of basic material for breeding new generation rice varieties. An analysis of rice production in the Russian Federation reveals the negative effect of the permanent increase in fuel prices: the self-cost of the grain rises and its competitive ability declines. The solution is seen in the development and introduction of energy-saving, ecologically safe rice-growing and harvesting methods, as well as in the breeding of new varieties. Varieties need to combine low energy requirements with adequate productivity, resistance to diseases and environment stress factors, and growth ability without the application of herbicides. The last 20 years of research at ARRRI have been dedicated to the release of such rice varieties.

Sample Assessment

The success of breeding activities largely depends on the availability of diverse initial material and the bank of crop genetic resources. Principal activities include the collection, safe-keeping, study, description and provision to rice breeders of samples. Rice collection in the Russian Federation was initiated in the 1920s by the specialists of the All-Union Institute of Plant Industry (VIR) under the guidance and with the direct participation of Mr N.I. Vavilov. The collection currently contains over 5 000 samples belonging to varieties and types that reach maturity under the climatic conditions in the Russian Federation. Special nomenclature of the *Oryza* L. genus has been established for rice evaluation using uniform assessment methods. The best rice samples in the world are included in the ARRRI working collection comprising over 3 000 viable samples of various pedigrees, including: mutants and polyploids; varieties that have been rejected or that did not pass the tests; selections from hybrid populations of senior generations; the best varieties in the world collection; and varieties commercially grown in Russia.

ARRRI and breeding centres in other rice-growing countries exchange material so as to have access to the latest achievements in world rice breeding for the creation of new varieties. Quarantine and introductory nurseries prevent the introduction of quarantine pests and diseases with rice seeds received from other countries.

The majority of varieties coming to Russia from other countries are late-maturing and—being as a rule native of the tropical zone—they are photosensitive and react to a 16-hour photoperiod (under conditions in Krasnodar) with an increased vegetation period of 150–160 days or more. The majority of samples do not reach heading and many of them are used only for hybridization under conditions of climatic chamber.

Foreign samples are often used as sources of traits such as short stem, long grain, high milling qualities, and resistance to pests, diseases and environmental stress factors; in return, ARRRI sends collection samples in response to the requests of foreign colleagues. Rice breeders in other countries evaluate the ARRRI samples and report back on the behaviour of the material under local conditions, providing interesting data, in particular with regard to the degree of resistance to pests and diseases under other ecological conditions.

Evaluation of Collection Material

Every sample in the ARRRI working collection undergoes complex evaluation of 40 traits to identify potential donors of economically valuable features. The samples are studied under field and vegetative conditions and in laboratory tests, with the participation of numerousspecialists: plant breeders, specialists in genetics, plant physiology, biochemistry and phytopathology from ARRRI and other research institutes.

In standard field experiments, the samples are evaluated and described according to their morphological traits, and their resistance to logging and shedding, and to pests and diseases is defined. In special experiments under provocation settings, the resistance to salinity, low temperatures, blast, aphids and rice leaf nematode is evaluated; the reaction to high nitrogen rates is also determined. Under laboratory conditions, the rice grain is assessed for its milling qualities: total milled rice and white rice, vitreousity, filminess, fracturing, grain size and form, and weight of 1 000 grains. The protein and amylose content is also defined.

Sources of early maturity, high-yielding ability, high grain quality, increased protein and amylose content, salt and cold tolerance, and resistance to pests and diseases are selected. An important feature is high yield, which is obtained by individual productivity of plants under optimal plant density. The working collection is constantly replenished with high production samples created in the course of breeding research. As productivity depends on the duration of the vegetation period, most sources of high productivity are medium- and late-maturing. Russian rice growing is one of the northernmost in the world; therefore special attention is paid to breeding rice with a short vegetation period. ARRRI breeders use early-maturing (up to 100 days of vegetation) and fast-maturing (100–110 days) rice samples to obtain early-maturing primary material. The collection includes over 250 samples.

Rice breeding for cold tolerance is based on valuable samples which combine tolerance to low temperatures with other economically valuable traits. Each ear ARRRI plant physiologists evaluate 100–150 collection samples for cold tolerance and the best are recommended for hybridization. The samples obtained though breeding for cold tolerance are added to the germplasm bank.

Large areas of rice systems are subject to increased salinity; breeding programmes for salt tolerance were therefore initiated. To define the sources of salt tolerance, as many as 400 samples from working and world collections are assessed each year. The best samples—combining high salt tolerance with other positive traits—are recommended for further breeding programmes.

Under Russian Federation conditions the most noxious disease in rice fields is blast, caused by fungus *Pyricularia oryzae* Cav. All collection samples are therefore evaluated for resistance to this disease and the best are used in hybridization.

In addition to high productivity and resistance to diseases and environmental stress factors, varieties should have vitreous grain and resistance to crushing—able to give maximum total milled rice and more economically efficient.

Protein content is an important indicator of rice alimentary quality. Protein is 98 percent assimilated by the human organism, and low protein rice varieties contain 7.33

percent protein (dry matter) while high-quality varieties up to 11.9 percent. ARRRI biochemists have selected samples combining increased protein content with maximum content of amylose.

Under Russian Federation conditions, the selected collection samples accumulate more protein and amylose than Krasnodarsky 424; they provide basic material for these traits and are recommended for breeding. The data obtained for each sample are registered in a special logbook and on catalogue cards and any hybridization includes data analysis for parent selection. ARRRI created the Rice Genetic Resources Database for: storing and processing of information on rice collection; automatic searches for sources of required traits; and the development of breeding programmes. Information about the collection is obtained rapidly and sample use is improved, in particular when there is a rare combination of individual traits.

Methods of Creation of Basic Material for Breeding New Rice Varieties

The principal method used at ARRRI for breeding basic material is intraspecific hybridization. Mutagenesis and biotechnological methods are widely adopted for production of new forms. Modern methods of hybridization include castration and flower pollination. The most efficient—while simple—method is pneumatic castration, for which a special device has been developed: the panicles are pollinated on the castration day and the parents are grown in climatic chambers under optimal thermal and photoperiod conditions. Hybridization programmes are carried out all year round.

The application of pneumatic castration and pollination are responsible for the considerable increase in hybrid grains: every year the ARRRI hybridization centre receives for 120–130 combinations (i.e. 40 000–50 000 flowers) as many as 20 000 hybrid grains. The average grain setting is

50–60 percent and for some combinations it reaches 90 percent; there is a high output of true first generation hybrids from 93.1 to 98.0 percent.

F_1 hybrids are grown in vegetation vessels: in winter in climatic chambers and in summer on vegetation plots. To overcome the problem of newly harvested seed germinating, the dormancy period is interrupted: F_1 hybrid grains are warmed in hot water (70 °C) for 7–10 minutes, then thermostat-controlled at 40 °C for 24 hours. Further germination then takes place at 28–30 °C (thermostat-controlled), reaching 95–98 percent. Hybrids of second and subsequent generations are evaluated and multiplied under field conditions. Rice breeders select elite plants with pre-planned parameters for the development of a breeding nursery.

Over the ensuing years the material obtained is studied according to generally accepted breeding methods. The evaluation process is performed by various scientists and specialists in phytopathology, entomology, plant physiology, biochemistry and rice grain milling quality. Before varieties are handed over for state evaluation, farming methods are developed and primary seed production is started. Every year ARRRI specialists issue two or three different varieties for state evaluation, including ones tolerant to environmental stress factors. The state register includes varieties with various characteristics: Sprint, Slavyanets and Leader (adapted for herbicide-free management systems); Kurchanka (salt tolerant); Viola (glutinose); and Snezhinka (long-grain). While they share high grain milling quality and tolerance to stress factors, they have very different morphological traits and biological properties, outlined below:

— *Sprint (fast-maturing) and Leader (medium-late-maturing)* are characterized by low requirements in terms of growing conditions and their fast growth during emergence when the shoots easily overcome a water layer of 20 cm; they are therefore recommended for cultivation without application of herbicides. Resistance to blast means that fungicide treatments are not necessary.

— *Slavyanets (medium-maturing)* is increasingly widely grown and belongs to universal varieties; it can be grown using any farming methods accepted in the farm.

— *Kurchanka (salt-tolerant)* is suitable for cultivation in rice fields with increased soil salinity, where it has achieved yields 500-600 kg/ha higher than other varieties. This variety is salt tolerant at both emergence and flowering, when other varieties are especially susceptible to salinity.

— *Snezhinka (long-grain, indica)* produces high-quality milled rice. Following state tests and production verification, the variety was included in the state register and admitted for commercial production. It requires no special methods during growth; it possesses effective blast resistance genes and therefore needs no chemical treatments. Processing the long grains remains a problem, however: the equipment in factories is adapted to milling short-grain varieties and so efforts are still required in this area.

— *Viola (glutinous)* was state commissioned for use in baby and diet food. The variety has been tested and patented in the Russian Federation.

5

Crop–livestock Systems in Conservation Agriculture

Combining ecological sustainability and economic viability while maintaining or improving agricultural productivity has long been a matter of concern for FAO, as has reducing negative environmental impacts. Conservation agriculture, which aims for zero tillage with the maintenance of a surface mulch to protect the soil surface and increase biological activity in the topsoil, is increasingly becoming recognized as an effective system of crop production that protects the soil from erosion while reducing the overall use of agrochemicals.

Vast areas of forest have been cleared in the tropical areas of Brazil for establishment of pastures that become unproductive once the native fertility of the soil is exhausted; this leads to yet more forest clearing for new pastures. However, rotating pastures with field crops and resowing is one of the most effective ways of maintaining them in a state of high productivity, thereby reducing the need for more clearing.

This chapter describes how pasture, fodder and livestock production have been integrated into conservation agriculture systems in Brazil's tropical zones. Integrated crop-livestock zero tillage systems (ICLZT) allow the sustainable production of high-yielding pasture without further deforestation; in this system, grazing livestock convert both pastures and crop residues into cash. The ability of pasture to build up the biological activity and physical quality of the soil is well known. The lessons learned in Brazil by farmers and scientists can provide valuable insights on what could be done in similar ecologies elsewhere, including in Africa where the details of management will be different, but the biological principles learned in Brazil could be a roadmap to more sustainable intensification of some major crop-livestock production systems.

Conservation Agriculture in the Brazilian Tropics

Brazil is a world leader in conservation agriculture (CA); initially the emphasis was on crop production through the now well-known principles of maintenance of a layer of crop residues on the surface, zero tillage (ZT) and crop rotations. Now Brazil is pioneering the integration of livestock production, grazed pasture and forage crops into CA, grazing being so managed as to provide adequate surface litter for the needs of ZT. A pasture phase in a rotation is renowned for building up soil organic matter (SOM) and improving soil structure. Pasture in ZT rotations with annual crops can be regenerated much more profitably and with less risk than the older systems where pastures were ploughed out before being resown.

Agriculture has had different emphases in the Amazon and Cerrado regions: beef, upland rice and small-scale dairying in the former; large-scale beef and soybeans in the latter, with maize, upland rice and cotton of secondary importance. Most clearing for pasture in the last 20 years has been on infertile soils, subject to rapidly falling stocking rates as initial fertility declines and little or no fertilizer is used. This has led to even more clearing to compensate for loss of carrying capacity.

About 55 percent of the Cerrado biome has been cleared as opposed to 16 percent of the Amazon biome. Land prices have risen; cattle farmers are selling to crop growers and moving to frontier areas to continue clearing. Land use intensification, through integrating ZT crops with livestock systems is the most viable route to preserve biodiversity on a large scale, provided that there are policy incentives for it and disincentives to clearing new land.

The Cerrado Biome

The descendants of seventeenth century mining pioneers grew subsistence crops (on the less than 7 percent of the area which had fertile soils) and reared beef cattle on the savannas, which were gradually impoverished through burning at the end of every dry season to generate regrowth.

Improved pasture choices in the Cerrado were dictated by soil fertility. On the few areas of fertile soils under forest (Nitisols), mostly colonized by 1970, Molasses Grass *(Melinis minutiflora)*, Jaraguá *(Hyparrhenia rufa)* and to a lesser extent, Colonial Guinea Grass *(Panicum maximum)* were sown. From 1972 these were replaced by *Brachiaria* spp. as fertility dwindled, or the pastures were renovated after a crop phase. In the latosols and quartz sands (Ferralsols and Arenosols) of the Cerrado the opportunity to mine native fertility by burning the sparse scrub is low, compared to Amazon forest. Pastures were established after one or two pioneer crops of rice, using very low fertilizer rates and minimal levels of lime, with little or no maintenance fertilizer. Sano *et al.* estimated that over 70 percent of Cerrado pastures were degraded.

Soybean production, the principal engine of development in the Cerrado, began significant growth from 1980. The Cerrado, with its gentle topography, wide interfluves, many tablelands, but essentially infertile soils, was opened for grain crops by migrant farmers from the southern states, mostly Rio Grande do Sul, using soil amendment technology from the 1960s. This technology was refined by the Brazilian Agricultural Research Corporation (EMBRAPA), but only came into its own with the advent of the first real tropical soybean cultivars "Cristalina" and "Doko" in the late 1970s.

Early development of large-scale arable cropping was based on subsidized agricultural credit. Conventional tillage was mainly with disc harrows which caused disc pans and surface capping, and progressively lower rainfall infiltration rates with erosion losses of up to 20 tonnes of topsoil/ha/year and rapid depletion of the already low soil organic matter, exacerbated by monocropping of soybeans. Contour banks, required as a precondition for rural credit, were inadequate for downpours of over 100 mm/hr and economic losses due to erosion gradually worsened as soil structure deteriorated. Zero tillage came into use when conventional tillage on infertile Cerrado soils was becoming unattractive, since soil amendments amounted to about 25 percent of direct costs and erosion losses were significant in terms of replanting and lost soil amendments. Since about 1990 a significant number of crop farmers have diversified into beef cattle, moving towards ICLZT systems, as they acquired capital and diversified their risks, or bought degraded pastures to expand soybean production.

On Cerrado latosols and quartz sands (Ferralsols and Arenosols), improved pastures were based on *Brachiaria decumbens* with some *B. humidicola* from 1972 onwards, usually undersown in upland rice, with low initial fertilizer levels. The *B. decumbens* cultivars were replaced from about 1990 by *B. brizantha* cv. Marandu and, to a lesser degree *B. ruziziensis.* Improved cultivars of *Andropogon gayanus, Paspalum atratum* and *Setaria anceps* and attempts to introduce pasture legumes had limited success. Predominantly extractive management led to progressive pasture degradation and impoverishment of cattle farmers, who tend to sell land to crop farmers and move to new frontiers. In rotation with crops, on amended Cerrado soils, new, less stemmy and less clumpy *Panicum maximum* cultivars (Tanzânia, Mombaça, Centenário and Vencedor) have supplanted the old Guinea grass since 1990.

The Amazon Biome

In the nineteen-seventies, farmers began to develop areas with fertile soils in the Amazon in southern Pará, Acre, central Rondônia and northern Tocantins states, where a forest-covered region with areas of podsolic soils of some initial fertility (Acrisols) supported Colonial Guinea Grass for up to 15 or 20 years. Guinea grass was short-lived on poorer soils and was replaced by *Brachiaria humidicola* and *B. decumbens,* and since about 1990, by *B. brizantha,* with a cycle of up to 10 years before requiring renovation. Large areas of short-lived Guinea grass, colonised by weeds and bush were abandoned due to high renovation costs, especially where there were unrotted stumps. Cattle expansion has pushed land clearing into rain forest on infertile soils with a much shorter depletion cycle.

Pasture development was mostly by incomers from south Brazil, spurred by subsidized credit. Small farmers (local and immigrant) in the Amazon opened forest land with subsistence crops, intersown with pasture in the second or third year as increased weed pressure made annual crops unprofitable. Pasture was established cheaply between the stumps. Large farmers either bought land with established pasture, or carried out more expensive mechanical clearing and broadcast pasture seed after a burn. Pastures were established without ploughing so there was some erosion control.

Where pastures are renovated with full land preparation, usually undersown in rice, the only conservation measure, when present, is contour banks, which can only be made once the stumps have rotted (about 20 years after forest clearance). Contour banks are only effective so long as the soil maintains a good infiltration capacity. This decreases after pasture renovation due to compaction from cattle hooves and reduced ground cover, through overgrazing, exacerbated by clumpy grasses like Colonial Guinea. Weeds from South Brazil were introduced in dirty pasture seed. Residual effects from former widespread use of Picloram herbicide for brush control in annual cropping have not been assessed on a regional scale.

As pasture quality declined, migrant small farmers were forced to intensify into milk production or depend on weaner-calf operations in conjunction with extractive crops like cassava. Few had the capital to invest in perennial crops on a significant scale, and extension services in the Amazon and access to credit have historically been weak for this sector. Many farmers are in settlement projects with 100 ha plots and have nearly exhausted the forest as a fertility source. Where small farmers adopted conventional tillage it was almost invariably with machinery hired from larger farmers, rarely employing contour banks.

Traditional populations used slash and burn on small 2-3 ha forest plots with a long fallow and were close to sustainable, but with marginal income.

History of Zero Tillage in Tropical Brazil

The first known tests with ZT in tropical Brazil were in Matão, São Paulo state (Atlantic forest biome) in the 1960s, with a sod-seeder for introducing legumes into pasture - however, this did not become general practice. Herbert Bartz began ZT on his farm in Rolândia, Paraná state, in 1972, just north of the Tropic of Capricorn, and has used it continuously thereafter. His experience is mainly suited to southern Brazil.

According to Landers, in 1979 small farmers in Rondônia were already jab planting beans *(Phaseolus vulgaris)* into rice straw after desiccation with paraquat as a means of weed control. From 1981 onwards, mechanized tropical ZT began to develop, with soybeans, maize and other crops in the Cerrado, reaching over 9 million ha in the tropical areas in 2004-05, with some encroachment on the transition forest in Northern Mato Grosso. Nearly all of this is in mechanized grain production. Today ZT is being further developed in Roraima, Rondônia, Amazonas and Pará.

The advent of tropical ICLZT technology in the mid 1990s enabled the sowing of grain crops directly into desiccated pastures, revolutionizing crop and pasture production through a synergy which benefited both. Since 1992 the Zero Tillage Farmers' Association for the Cerrado Region (APDC) has actively promoted ZT and, latterly, has been in the forefront of promoting the ICLZT technology. Zero tillage is now the dominant production system for annual crops in Brazil and is increasing rapidly in importance for pasture renovation and establishment of perennial crops. ICLZT has shown tremendous potential for land use intensification, reducing the rate of clearing of native vegetation.

Conservation Agriculture

The term "conservation agriculture" was adopted during the First World Congress on Conservation Agriculture, Madrid 2001, organized by FAO and the European Conservation Agriculture Federation.

The distinguishing and most important element of CA is that crop residues remain on the surface, including those of cover crops (green manures). The many functions of this surface organic layer, slowly decomposing to humus, transform an untilled soil into a living, dynamic system which must be managed to maximize these functions and derive the benefits which they confer, resulting in increased SOM, greater P availability and faster breakdown of agricultural chemicals. Conservation agriculture replicates the closed cycle of nutrients and surface litter of a mature rain forest, bringing highly productive farming into harmony with nature. Water-holding capacity increases in proportion to SOM and improved soil structure through an increase in water-stable aggregates; water economy also improves since mulch reduces evaporation losses. Stone and Moreira measured reductions in irrigation demand of over 30 percent with an erect cultivar of *Phaseolus vulgaris* planted in a thick mulch.

To plant in permanent residue cover it was necessary to develop specialized planters and drills that cause minimal disturbance. These are now available worldwide in manual, animal-drawn and mechanized models. The sowing mechanism consists of a trash disc which cuts the crop residue and allows the fertilizer and seed to be placed beneath it with minimum soil disturbance, leaving the soil protected against rain and sun.

Contrary to early predictions, once all plough-pans are removed (as a pre-condition to adoption) and movement of lorries and heavy grain trailers is restricted to roadways, ZT maintains soil structure at depth by the preservation of macropores derived from old root holes and the galleries and burrows of soil mesofauna, such as earthworms, beetles and their larvae. Basic infiltration rates after two hours under these conditions have been measured at 120 mm/hr, far higher than even tropical rainfall intensity over the same period; without such macropores this drops to 30 mm/hour or even 20 mm/hr when soil is compacted. The macropores thus provide effective erosion control under ZT; long term ZT farmers are removing their contour banks because they no longer need them when they have infiltration rates higher than rainfall intensity and a good residue cover protecting the soil. An average of 30 erosion experiments in Brazil showed a reduction of 79 percent in soil losses under ZT compared to conventional tillage (5.6 tonnes/ha/annum under ZT and 23.3 tonnes/ha/annum with conventional tillage, while soil regeneration is estimated at 10 tonnes/ha/annum).

How does Conservation Agriculture Work?

The first operation of the agricultural year is desiccation of cover crops and/or weeds with non-selective contact herbicides of negligible environmental impact. Thereafter, specialized planters, or drills, cut the crop residue with trash discs and slot the fertilizer and seed into the soil with minimal residue disturbance. The only subsequent operations are application of agricultural inputs (if required) and harvesting. Combine harvesters are fitted with straw spreaders or choppers to give even cover of crop residues. Conservation agriculture has engendered a more responsible attitude towards the environment, regarding nature as an ally, and not as a foe. The CA farmer combines crop rotations and maintenance of surface residue with cover crops, integrated management of weeds, pests and diseases, rational fertilizer practices, integration of crop and livestock enterprises, watershed management and other environmental concerns. This confers sustainability on the system and assures zero or low levels of chemical residues in agricultural products, well within prescribed legal limits. Landers estimated values for ZT and ICLZT environmental impacts in Brazil.

Integrated Crop-Livestock Systems with Zero Tillage

Zero tillage has facilitated the development of ICLZT in the Cerrado biome, with wider application, in principle, in the Amazon, Atlantic Forest and Caatinga biomes. Arable CA systems encompass a whole farm, with both crops and livestock. Integration of beef fattening or rearing with annual crops started as diversification under conventional tillage. The new tropical ZT technology eliminated the erosion risk from ploughing, gave higher crop yields and tripled pasture carrying capacity while reducing costs.

Farmers have encouraged technology development, facilitated trials on their farms and often used technology ahead of research results. The backup of wide-spectrum and interdisciplinary field research (governmental and private) has resulted in credible long-term farm results, specialised training and extension programmes.

A new era for tropical CA began in the 1990s, triggered by ICLZT, which gave the potential to increase farm profitability and reduce clearing of native vegetation through land use intensification. A crop phase is the most cost-effective means of maintaining highly productive pastures. Many cattle rearers lack cropping skills - in which case a long term rental agreement with a crop farmer, guaranteeing winter grazing on the stubble until the first crop land reverts to pasture, can satisfy both sides, especially if *Brachiaria* spp. are undersown in the summer crop. Crop farmers diversify into cattle more readily, not least because, in good years, they have capital to invest and wish to spread their financial risk; ranchers are more averse to ICLZT because of (generally unfounded) fears of losing carrying capacity.

The benefits of ICLZT are higher than for crop and livestock systems conducted separately, illustrating that the benefits discussed above translate into profits.

DISSEMINATION OF ICLZT TECHNOLOGY

A series of seminars, demonstrations, farmer technology meetings and training courses involving ICLZT have been carried out since 1997, mostly initiated by the private sector, more recently with active participation of EMBRAPA, which now leads these efforts. State extension agencies, which primarily serve family farmers, have been active in ICLZT in the last few years. In 2003 the Rice and Beans centre of EMBRAPA published a complete manual of results on ICLZT and in 2005, the Ministry of Agriculture, Livestock and Food Supply launched a programme specifically to promote ZT, with training, demonstrations and research. In 2004 APDC initiated its "Stewards of Our Water" project with Petrobras and in 2006 a joint APDC-National Water Agency (ANA) project was launched, using ZT and ICLZT as the principal instruments to improve on-farm water conservation. The "Pure Oil from Soya" project financed by Stichting DÓEN from the Netherlands aims to reduce chemical use in soybean production. These projects obtain secondary financial support from the private sector, state and municipal governments.

Livestock and Annual Crop Production in Wet-dry and Humid-tropical Brazil

Livestock Type

Beef cattle breeds in tropical Brazil are dominated by the Ongole zebu from India, which has been the subject of selective breeding since the first importations in 1870. The Nelore is the principal breed of Indian origin, which has other minor derivatives. Other breeds from the sub-continent are losing popularity as they are upgraded with superior Nelore bulls. *Bos taurus* breeds dating from the colonisation period and Santa Gertrudis *(B. taurus* x *B. indicus)* are of minor importance. European beef breeds are not adapted to the tropics. For fattening, crosses of Nelore or other zebus and European breeds are favoured; artificial insemination is also used. Cows from these crosses come into heat at about

24 months, as opposed to 30 months for pure Nelore. Magnabosco *et al,* demonstrated the effect of genetic selection for performance on pasture: superior genotype animals showed a 26 percent improvement in weight gain over poorer ones.

Most of Brazil's milk comes from Holstein-Friesian cattle and their crosses, of which the Girolanda is the most popular. At tropical elevations over 600 metres, pure Holstein-Friesian perform well when given optimal conditions of feed and shade. Progeny testing and selective breeding programmes are advanced, as are specialised suppliers of semen and embryo transplant services. Holsteins were predominantly of North American origin, but recent imports of stock and semen from New Zealand and other countries are genetic types selected for performance on pasture. There are a few specialized Jersey herds.

Water buffalo thrive where they have access to water. Except for Pará's Amazon flood-lands the water buffalo population is low in the rest of Brazil. They are now expanding from the Amazon lowlands to the lower tropics, due to their hardiness and adaptation to poorer feed. There is an excellent example of almost subtropical ICLZT with water buffalo in Rolândia-PR, of many years standing, where biomass in excess of 6 tonnes/ha is made into hay for winter feed on permanent pasture.

Herd Size and Performance

Brazil's cattle herd grew from 153 million in 1995-1996 to 205 million in 2004, an annual growth rate of 3.7 percent. Of these, 60 percent are in the humid and wet-dry tropical states comprising the Amazon and Cerrado biomes. The Cerrado figure is estimated since IBGE figures are by state, not biome. The importance of the Cerrado in beef production is evident. About 60 percent of Brazil's 99 million ha of planted pastures are in wet-dry and humid tropical regions.

The southern winter offsets the dry season in the Cerrado, and most of Brazil's herd is in these two regions, so these figures are considered representative for the wet-dry and humid tropics. The livestock sector in tropical Brazil is at the end of an extractive phase where the solution has traditionally been to expand onto newly cleared land. From about 1980, expansion was mainly on dystrophic latosols (Ferralsols) or quartz sands

(Arenosols), which required liming and had little inherent fertility. Pasture establishment under rice, seldom using more than 300 kg/ha of NPK 4.14.8, was common. Stocking rates fell fast and clearing continued to compensate for both loss of carrying capacity and herd expansion.

Background for ICLZT

Integration of beef fattening or rearing into annual crop systems began as a hedge against low soybean prices, but with the advent of the new tropical ZT technology, its multiple benefits to both crops and pastures soon became evident. Dry season pasture production determines a farm's year-round carrying capacity and beef prices rise in this period. In this publication "carrying capacity" refers to the dry season stocking rate.

About 85 percent of pastures in the Cerrado are *Brachiaria* with somewhat less in the Amazon biome, while *Panicum maximum* cultivars make up less than 5 percent in the Cerrado. Introduction of *Brachiaria* in the Cerrado raised average carrying capacity from 0.3 AU/ha to 1.0 AU/ha. *Brachiaria* pastures on Cerrado latosols can last up to 5 years without further fertilizer; Guinea grass cultivars give higher carrying capacity on more fertile soils, but decline rapidly with loss of fertility. The winter carrying capacity of native pasture is between 0.1-0.05 ha/AU with annual burning to generate regrowth at the height of the dry season. Under this regime mortality was high and steers lost weight badly over the dry season, reaching slaughter (220-240 kg carcass weight) in three to four years.

Extensive beef producers have been slow to invest in management technology, (e.g. fertilizing pastures, electric fencing, pasture legumes and integrated crop-livestock rotations) but have rapidly taken up new grasses, such as *Brachiaria brizantha* and new cultivars of *Panicum maximum,* which show a very high benefit/cost ratio. Extractive management led to gradual decapitalization of beef farmers, which, together with resistance to change, has impeded the modernization of the industry. Growing maize and sorghum for silage is quite widespread and dry season fattening pens have gradually increased since the early 1970s. Since about 1990, there has been a significant diversification of crop farmers into beef, moving towards ICLZT.

Vaccination against foot-and-mouth disease, control of ectoparasites and ear tagging are obligatory in beef-exporting states and the rule in the other states, except for ear tagging. Mechanical milking is increasing rapidly, although most small producers still hand-milk; silage or green chop are used extensively in winter and concentrate feed is based on maize and soybean meal with additives. Small farmers are traditionally milk producers on improved pastures; in the South of Pará there is a concentration of small milk producers, currently in crisis due to falling carrying capacities and lack of access to new land.

Available pasture technology permits profitable intensive cattle operations. Electric fencing facilitates rotational grazing and the adoption of ICLZT is a low-cost way to turn residual fertility from the crop phase into profit. Fertilizing pastures is considered uneconomic at prevailing beef prices.

The Process of Pasture Degradation

Inadequate replacement of nutrients and overstocking are the chief causes of pasture degradation in the Cerrado. Before going into an ICLZT system the degree of pasture degradation must be taken into account, to decide whether full land preparation is required. Besides fertility decline and overstocking, brush invasion and large paddock size are the major limitations to pasture productivity in the Amazon.

Continuous degradation through extractive management brings successive reductions in carrying capacity. These begin with N and P deficiencies, and end up in the loss of soil cover and consequent soil degradation through compaction and erosion. Overgrazing is the most common form of poor pasture management. Ideal management seeks to keep the pasture with full ground cover and always in the productive phase, by adjusting stocking rates or rest periods to avoid over-grazing, and adding fertilizer and lime when necessary. Short high stocking periods in the maintenance phase are tolerable.

In the Amazonian pasture areas of southern Pará the 2-3 month dry season and better quality Guinea grass carried 1 to 2 AU/ha over the first five years, depending on initial fertility. When it dropped to about 0.5 AU/ha, due to fertility decline or bush encroachment, the pasture was renovated with *Brachiaria* spp., with carrying capacity ranging from 1.5 to 2 AU/ha in its first five years. Thus, *B. brizantha* or *B. humidícola* replaced Guinea grass as fertility was mined, restoring profitability with a second extractive phase.

Principal Integrated Zero Tillage Crop-livestock Systems

Zero tillage is not sustainable without crop rotation, always with a different crop in the same season of the subsequent year. Rotations comprise a number of single-year crop successions. In tropical Brazil a second crop may be sown in the second half of the rainy season: where rainfall is inadequate a fallow is used. Examples of typical rotations are found below and others are detailed in the case studies.

Soybean-millet is a common crop succession in one year, but soybean-millet followed by soybean-millet is not a rotation, it is a crop succession repeated. The greater the number and the higher the diversity of crops and genera involved in a rotation, the higher the biodiversity and the greater the potential for biological control of diseases, pests and weeds, through cutting the build-up of inoculum or populations; even nematodes can be efficiently controlled by some pasture grasses and leguminous cover crops. A pasture phase in a rotation builds up SOM, improves soil structure, nutrient and water availability, and allows reductions in the levels of pesticides and fertilizers used. High SOM and crop residue levels reduce the potential pollution of aquifers and surface water by acting as chemical and physical buffers.

Crop successions within the rotation must be balanced so as to maintain an average of over 6 tonnes/ha of dry matter in crop residues. Production of adequate biomass to build up SOM and maintain profitability is helped by a pasture phase. Cover crops can assist when late or winter rains are adequate. Sá *et al.* demonstrated increases of root mass of between 11 and 76 percent at 5 and 10 tonnes DM/ha of residues with a positive regression on hybrid maize yields.

Biological activity in the soil is increased under ZT The most important discovery in this area, using *Brachiaria* spp. as a cover crop, has been the reduced incidence of soil pathogens, as shown by Da Costa and Rava, who measured reductions in *Rhizoctonia (R. solani)* and White Mould *(Sclerotinia)* infections of 75 percent and also in *Fusarium solani*. This principle has revolutionized rainfed and irrigated Phaseolus bean production when planted into desiccated *Brachiaria,* sometimes leading to farm yields of over 3 000 kg/ha . Dos Santos analysed different systems of ICLZT in terms of weight gain (in beef) and financial returns to the beef enterprise. The introduction of a fast-establishing annual (hybrid forage sorghum), gives grazing 20 days earlier compared to conventional establishment of *Panicum maximum.* Alternatively, a mixture of pearl millet *(Pennisetum americanum),* or finger millet, *(Eleusine coracana)* with *P. maximum* can achieve the same result.

There is a direct correlation between the amount of crop residue dry matter and the yield of the following crop, as shown by Séguy and Bouzinac. The farmer must know this to gauge the opportunity cost of the pasture consumed, but this requires a level of cost analysis beyond that of most farmers. Pasture renovation with ZT can be done without a crop phase, by desiccation of the old pasture, re-seeding with fertilizer and liming, using part of the technology described on the following pages.

Systems Typology

Livestock systems in ICLZT are independent of the crops grown; the main ones are:

(i) Fattening steers to slaughter weight;

(ii) Rearing operations selling steers to (i);

(iii) Combined rearing and fattening operations;

(iv) Cow calf operations selling weaners to (ii) and (iii);

(v) Breeding-to-slaughter integrated operations;

(vi) Intensive dairy operations;

(vii) Small, non-intensive, dairy operations.

All may or may not include supplementary feed with silage, cut fodder and concentrates. In beef-from-pasture operations this is confined to the dry season and creep (supplemental) feeding of calves at high management levels. Use of salt or mineral mixes is general, even in the most extensive systems, and mineral supplements usually include protein or urea to improve digestibility of dry season forage. Feedlot fattening can be grafted on to (i), (iii) and (iv), considerably increasing intensity of land use. Pure feedlot fattening is not discussed here as it is an independent alternative to fattening on pasture.

There is no over-arching crop system typology. The chief distinction is between rainfed and irrigated cropping, which is rarely without a rainfed area. Few farmers change their basic cropping pattern when adopting ICLZT, although they may undersow maize or oversow soybeans with grasses or grow earlier varieties to allow greater winter pasture production as a second crop. Some advanced farmers irrigate intensively-grazed pasture. Cattle farmers adopting ZT cropping (a lesser trend) usually choose soybeans or cotton as their main cash crop. The latter is late-sown in December, which allows time for early bite grazing at the onset of the rains with pearl millet, finger millet or forage sorghum.

Basic ICLZT systems have a crop and a pasture phase in rotation and comprise:

(i) Winter stubble grazing on summer cropland;

(ii) Summer crops with winter pastures;

(iii) Summer crop plus second crop plus stubble grazing in winter;

(iv) Crop production for feed supplement, usually a minor area within another system;

(v) Some combination of these.

Common Rotations

Soybeans are often mono-cropped for several years during

the low disease-weed-pest pressure period after land clearing and are universally dominant in rainfed farming, with maize in rotation, when practised, every two to four years. Soybeans do not return enough biomass to the system for long term sustainability While physical soil degradation has been severe under conventional soybean cultivation it has been hardly noticeable under ZT, leading many farmers to disregard a pillar of CA, crop rotation.

Maize is essential for adequate biomass generation in a CA rotation, but in frontier areas the market is weak; higher transport costs per unit value and higher storage requirements (both double those of soybeans) severely limit its use in rotations. In such regions rice is often grown as a pioneer crop after clearing, or in pasture renovation. New rice cultivars can exceed 5 tonnes/ha. Cotton and Phaseolus beans have the advantage of late sowing, allowing more grazing on winter pasture. In the western Cerrado and in Amazonia a second crop of maize can be highly profitable. Alternatives to maize as a second crop in lower rainfall areas are pearl millet *(Pennisetum americanum),* sorghum *(Sorghum bicolor)* and finger millet *(Eleusine coracana),* or pasture grasses sown in association with the main or second crop.

A common starting point for adoption of ICLZT would be a 10 to 20 year-old degraded pasture. The most generalized ICLZT rotation would be: Yrs 1, 2 and 3 soybeans in summer and stubble grazing in winter; Yr 4. maize, undersown with *Brachiaria brizantha* or *B. ruziziensis;* years 5-7 or 5-8 pasture.

The most sophisticated rotation would be: Yr 1 cotton-winter fallow, Yr 2 soybeans-second crop maize, Yr 3 soybeans-second crop maize undersown with finger millet and *Panicum maximum,* Yr 4 one-year *P. maximum* ley. This maximises the stocking rate on the pasture up to 4-5 AU/ha, resulting in 75 percent of the farm in crops.

The simplest rotation would be that of Case Study 5, a predominantly cattle operation, with: 6 year-old pasture converted to Yr 1 maize for silage and/or grain, undersown with grass, Yr 2-7 pasture (enriched, or not, with pigeon pea).

Crop Successions used as Building Blocks for Rotations

Crop successions, which are the building blocks for any desired rotation, can be classified as follows:

- Crop establishment in degraded pastures;
- Pasture establishment in, or immediately following, an annual crop (for permanent or temporary pasture). This could include dual use of a cover crop both for grazing and subsequent soil cover;
- Use of crop areas for silage or green chop production as supplement for cattle;
- Enrichment of existing pastures with legumes. This is practically limited to sowing pigeon pea directly into existing pasture;
- Sowing annual crops into a permanent grass or legume sward;
- Opportunity grazing of crop stubble in the dry season.

Summaries of the Ten Main ICLZTTechnologies

1. Crop Establishment in Degraded Pastures

Excessive impediments to mechanized cultivation, or toxic aluminium levels in the 0-20 cm layer, may require full land preparation to remove irregularities and incorporate lime. When necessary, operations should be carried out towards the end of the rains and an annual fodder sown immediately to provide enough winter grazing to more than compensate for loss of the old pasture. This removes the chief concern of the rancher (i.e. lack of winter forage). These fodders must be grazed to suppress flowering or they will mature and die back. If they set seed, it will fall to the ground and germinate with the first rains. Either from regrowth or seed, ground cover can be generated and ready for desiccation prior to sowing the annual crop. In the case of subsoil deficiency in calcium or sulphur deficiency, gypsum can be applied along with lime, whether incorporated or not.

If no impediments exist the degraded pasture is desiccated at the beginning of the rains with a systemic, non-selective herbicide (glyphosate, or ammonium gluphosinate). Specialized ZT planters place seed and fertilizer below the residue. A narrow in-row tine or special fertilizer boot behind the cutting disc is necessary in old, degraded pasture to break the compacted surface layer, allowing deep fertilizer placement and penetration of crop roots.

Where there is no compacted layer or excessive residue to clog the tines, offset double discs of different diameters for both seed and fertilizer are used. Thereafter selective herbicides are used *ad hoc* in the crop. A one-year pasture ley maximizes the proportion of crops in the rotation for a given herd size, which is an option when crop prices are high. When soil conditions are adequate any annual crop can be sown with this technology - soybeans, phaseolus beans, maize, cotton etc. There is a special bonus with phaseolus beans because of the low level of soil disease and nematodes after desiccating *Brachiaria*.

2. Establishing pasture in annual crops

The most common options for establishing pasture in annual crops are: undersowing intersowing and oversowing. Rainfall availability is greatest with undersowing and least with sowing after harvest. CAT Uberlândia has shown no difference in maize yield when 9 kg/ha of *Brachiaria ruziziensis* seed was sown at the two and four leaf stages of maize, but the two leaf sowing gave quicker establishment and allowed earlier grazing. Kluthcouski *et al.* affirm that *Brachiaria* undersown in maize can be shaded to the point of not requiring herbicide to check it (tropical grasses are C4 plants, hence shade intolerant). For manual systems, a jab planter can be used, mixing seed with fertilizer or some inert granular material, but if the stand is not uniform shading will be inadequate and some yield loss will occur from competition from *Brachiaria* in the unshaded areas.

Oversowing can gain up to three weeks on same-day sowing after harvest, which is crucial because the amount of rainfall after germination is correlated with establishment and dry matter yield. However, conditions for establishment with oversowing are not always adequate and this risk must be accepted against the potential benefit and low cost factors. Intersowing is carried out by planting in the inter-row of the crop after emergence.

Maize undersown or intersown with *Brachiaria* is the most important method for establishing pastures, and may have nurse crops other than maize e.g. rice. Grass may be undersown, mixed with fertilizer at sowing, dropped onto the row in front of the trash disc from additional seed hoppers or by shallow drilling two rows in the inter-row at the 2-4 leaf stage of the maize. Excessive grass growth can be checked with sublethal doses of Nicosulfuron of 8-12 gm a. i./ha. Experiments in several locations and years showed little or no significant difference in maize or sorghum yields.

In upland rice, undersowing has been common under conventional tillage and works well with ZT when the grass is sown in the rice row to control shading. Although some yield reduction of rice has been observed - the same depths of sowing should be observed as for maize. undersowing is possible in soybeans: reduced post-emergent applications of haloxyfop-methyl are promising and Cobucci and Portela recommend one quarter of the full application to check *Brachiaria* spp., but this technology is not well field-tested and results will be very dependent on the metabolic activity of the grass.

3. Sowing pasture after early harvest

Usually, but not always in the Cerrado, there will be sufficient rain to establish a pasture up to late. The ideal method for sowing is with a ZT drill in closely-spaced rows (approx. 15 cm), straight after harvest (i.e. behind the combine on the same day). Drilling is preferred, but most farmers do not have a drill and broadcast the seed with a fertilizer spreader, covering it by running a closed levelling disc harrow over the stubble. This low-cost method leaves about 40-60 percent cover, but it oxidises organic matter and increases the evaporation of precious moisture. In addition, it only has a moderate chance of success after February (i.e. it is best after an early harvest, but then the farmer's choice would be for a second cash crop, so pasture establishment with this system will always be in moisture-short situations). Finally, care must be taken to minimize soil and residue disturbance - ideally, over 70 percent cover should be maintained to guarantee erosion control.

An innovative São Paulo farmer's solution to this problem was the adaptation of a levelling disc, welding angle iron bars across old, worn, discs, which leaves the cover practically intact. A chain harrow, or light anchor chain, between two tractors, are superior solutions and faster operations, incorporating the broadcast seed into the crop residue with minimai disturbance. Desiccation may be unnecessary if the weed burden is very low. Previous history of the field, plus spot observations should determine this option - persistent weeds, such as *Sida* spp., *Cassia rotundifolia, Commelina* spp. and *Digitaria insularis* should not be present. As much starter fertilizer at planting as the farmer can afford (emphasizing P) will assist in promoting good weed suppression and providing early bite. The amount of winter pasture in the first year will be governed by rainfall and overgrazing should be avoided.

4. Grass oversown in soybeans or maize

Oversowing can be done with a fertilizer broadcaster or by aerial seeding after field-testing to check the required overlap. Commonly-used forages in soybeans at full grain maturity are: pearl and finger millets and *Brachiaria* spp., but Guinea and Molasses grasses have also shown some success. This system is cheap (just the seed plus application), offset by high risk of failure if rain does not follow shortly after seeding; the Cerrado and Amazon transition forest regions experience short dry spells of ten days or more during January-March when soybeans are maturing. Oversowing into maize increases the risk since there is no leaf-fall to cover the seed; it should be carried out at the beginning of canopy senescence, when more light penetrates to ground level. With any main crop, the earlier it is sown and the shorter its cycle, the greater amount of rain will fall on the second (oversown) crop. Rarely will the conditions exist for oversowing of the whole cropped area. However, the cash value of additional dry season grazing is very high. Beware of ants, beetles, other insects or birds reducing establishment by removing seed - an increase in seed rate may be required. Seed dressing with insecticide in this sowing mode is dangerous for birds and small animals, and should be avoided.

5. Grass regenerating during the first crop after ZT planting of a crop in old pasture

As the crop canopy thins, seeds will germinate from the surface seed bank. Under rainfed conditions *Brachiaria* spp. will not be daylength-triggered to flower before it is time to desiccate for a second crop, so the seed bank will dwindle. The decision to leave for winter pasture or desiccate post-harvest with a half dose of systemic desiccant will depend on the grass stand and the weed load. Dense stands of *Brachiaria* are good at weed suppression.

6. Planting forages on crop land for silage, green chop, dry season grazing or as a cover crop

Forage sorghum, pearl millet or finger millet can be drilled with ZT as a second rainfed crop for green chop forage, silage, grazing or cover crop biomass. Sowing in January, or first half of February, will maximise rainfed dry-matter production, but this is seldom done, since the forage would compete with an economic second crop. Drought-resistant crops are usually sown after March 1 st. Drilling is the preferred method for millets. Because of high residual nutrient levels from the preceding crop, little, if any, fertilizer is used.

Irrigated silage and green chop, if possible, should be scheduled away from competing with beans or other high value crops in the dry winter, when these are grown - there is inevitably a conflict with green-chop for fattening yards whose demand is constant. Seed rates for forage sorghum are between 8 and 12 kg/ha in 50-70 cm rows, for pearl millet 15-20 kg/ha and finger millet 10-15 kg/ha, both in 15-25 cm rows. Seed depth for sorghum should be 3-4 cm and <2 cm for millet. In high fertility conditions under irrigation, for silage production, specialised sorghum hybrids can out-yield maize (up to 70 tonnes of silage/ha), at reduced fertilizer rates, while maize reaches about 45 tons/ha under the same conditions.

Some intensive systems have been producing two crops plus dry-season grazing: Séguy and Bouzinac report soybean yields of 4 000-4 600 kg/ha, 1 500-3 500 kg/ha of second crop grain plus 1–1.5 AU/ha for 90 days in the dry season with a mixture of the three forages. This gave a total gross margin for the annual succession of US$ 150-350/ha.

7. Pasture renovation with forages sown jointly with grasses, for early grazing

A specially-designed ZT combination drill has three hoppers (grass seed, forage seed and fertilizer). A set of trash discs cut through crop residue, followed by a narrow fertilizer boot with scarifier tine in the row (to cut through the 5–10 cm surface-compacted zone in the old pasture and place the fertilizer about 15 cm deep). This is followed by alternate sets of offset double disc furrow openers with different size discs for the grass and the forage sorghum, respectively, at about 15–18 cm spacing with open clodbusting rollers behind.

8. Pigeon pea sown into existing pasture to improve winter grazing quality

For sowing pigeon pea into existing pasture, the seeder-scarifier is used or a normal ZT planter is used with a fertilizer knife to break the surface-compacted layer, and not a double disc. Fertilizer (especially phosphate) applied in planting will also renovate the grass and the pigeon pea can fix up to 150 kg/ha of N, also improving grass vigour. The extra protein of the legume allows cattle to utilize poor quality winter grass efficiently. Pigeon pea seed rate should be 50 kg/ha and planting depth 4-5 cm. This technology can be adapted for *Leucaena leucocephala* and other legumes, although there is little local experience. *Leucaena* suffers from severe leaf drop in the dry season, and did not fulfil early promise.

9. Sowing perennial legumes into maize

This technology has good potential, but requires more development of control mechanisms to check established perennial legumes.A forage legume is established by undersowing in maize either using a pelleted insecticide applicator coupled behind the seed discs or by mixing the seed with phosphatic fertilizer and directing the fertilizer spout towards the soil falling back in the furrow after the seed disc. This implies pre-planting fertilization for the maize and topdressing of K and N, or a combination planter with three hoppers. The most compatible legumes were Siratro *(Macroptilium atropurpureum)*, followed by centro *(Centrosema pubescens)*, but kudzu *(Pueraria phaseoloid.es)* and glycine *(Neonotonia wightii)* were more competitive.

For silage, this system is excellent, leaving a full soil cover and giving high quality winter grazing which can be used as a protein bank. Also, the legume and surface litter will reduce surface compaction from machinery A double straight disc between the rows, at about 10 cm from the maize row, can cut the tendrils and prevent the legume from climbing on the maize.

10. Sowing soybeans in a permanent grass sward

This technology nas been tested in Mato Grosso; the idea is not to kill the grass, but to arrest growth up to soybean grain maturity. A Tifton cultivar of *Cynodon dactylon* was used with reported soybean yields of over 3 000 kg/ha. The grass, which can be grazed in the dry season, helps to suppress annual weeds and provides biomass for soil cover. The same authors obtained 90 days winter grazing for 1–1.5 AU/ha with this system and a total gross margin of US$ 200-400 per ha for the annual succession.

Opportunistic Grazing of Stubble in the Dry Season

Pigeon pea undersown in maize for stubble grazing

This is a traditional practice reported by Von Schaffhausen. Pigeon pea seed is mixed with the fertilizer and deep sowing

(10-15 cm) delays its emergence in the maize row, where it remains shaded until maize senescence. Manual jab planting in the row would follow planting of maize after 30 days. In the second year the pigeon pea should be rotary chopped before sowing and the regrowth sprayed with a systemic desiccant herbicide or a split dose of paraquat (1.5 and 1.5 L/ha of commercial product). If left to grow in the second year, it will need to be regularly grazed or it will become too woody for a combine cutter-bar or easy manual eradication. However, it could then still serve as a source of firewood or fodder for a small farmer.

Grazing stubble in the dry season

Nonavailability of fences and watering facilities can limit the use of stover by cattle. Electric fences are cheap, with an investment cost in Brazil of about US$ 80-100 per kilometre. The ratio of stubble area to permanent pasture area is crucial to performance in this system. The greater this is, the more positive the impact on farm total carrying capacity.

Pasture Grasses

Pasture grasses need to be selected according to the soil fertility level and internal soil drainage, a problem mostly confined to wetlands. In spite of the availability of a wide range of tested pasture legumes, there has been negligible uptake of mixed grass-legume pastures in Brazil, due to:

(i) real or imaginary difficulties in grazing management of these pastures;

(ii) excessive cost of seed, due to inefficient establishment of the legume and high market prices because of low demand; and

(iii) lack of efficient establishment systems for mixed pastures, which could start by undersowing legumes in maize and direct drilling the grass at the onset of the following rains, when the legume would have the advantage of an established root system and be able to compete with the grass coming up from seed. Landers indicated how to establish pasture legumes with maize using alachlor as a selective herbicide for both. On-farm research is required on these aspects.

The traditional Molasses and Jaraguá grasses *(Melinis minutiflora* and *Hyparrhenia rufa),* adapted to poorer soils, have been out-performed by the Brachiarias and their use is now rare. Quinn *et al.* showed the inferior performance of Molasses grass compared to Colonial Guinea Grass in all situations, and inferior performance of Jaraguá in high fertility situations, but holding its own under low fertility.

Cover Crops for Grazing

Pearl Millet, *Pennisetum americanum.* This can be used purely for biomass production, for high-quality grazing, for hay or for seed, although rarely yielding more than 500 kg/ha of grain in late autumn sowings, due to rainfall limitations; the protein content of its grain is almost twice that of maize. Harvest is by combine harvester (by hand on small areas) and most farmers produce their own seed. Breeding programmes have still not released hybrids; these should make millet a good option as a grain crop for very late summer plantings as it can extract nutrients from depth and can be sown on residual fertility, with just a small top-dressing of N.

Pearl millet can be a host for army worm, but it controls nematodes well and its residues apparently encourage less slugs in irrigated conditions than those of maize. Millet is a slow starter, accelerating development from about 30 days after sowing. After the first month, in spring it can produce over 100 kg DM/day, but much less as days get shorter. Millet can smother weeds sufficiently to eliminate, or reduce, post-emergent herbicide use and has a similar effect in autumn once a good cover is achieved. Under grazing it should not be allowed to initiate flowering, in order to continue in vegetative mode.

Black Oat, *Avena strigosa.* This is not widespread in the tropics and is limited to latitudes close to the tropic of Capricorn (São Paulo, Paraná and Mato Grosso do Sul states) or altitudes above 1 000 m, as it does not take well to heat. It dies back in summer and only requires rolling, obviating pre-plant desiccation and often eliminating the need for post-emergence herbicides, besides acting as a good recycler of nutrients and incorporator of calcium in depth by translocation to its roots. When grazed it should not be allowed to initiate flowering to maintain it in the vegetative phase.

Forage Radish, *Raphanus sativus.* This crop has a deep tap root and is a good recycler of N and other nutrients and has excellent forage value. It is more a subtropical crop and performs best near the tropic of Cancer, but has been used successfully at 16° S and 1 000 m altitude. Its small seed requires care in sowing. This crop is a good precursor for maize. Forage radish's thick. carrot-like roots act as biological sub-soiling agents. It may advantageously be mixed with other cover crops, especially black oats, giving complementary cover, with a different growth cycle. The Instituto Agronômico do Paraná has released improved varieties of both black oats and forage radish.

Cut Forage and Silage Crops

Commercial hybrids of maize and forage sorghum or sorghum x Sudan Grass hybrids are the principal forages used for silage, and some open-pollinated varieties from the Guinea coast of Africa have shown promise. Resprouting ability after the dry season is an important quality to evaluate in sorghums. Sugar cane, elephant grass and recently, *Bracniaria brizantha* cv. Marandu are the most important plants for cut forage. The importance of the latter is that it can fit into a rotation as a catch crop, whereas sugar cane and elephant grass occupy the same land for several years of ratooning. Kepler recommends cutting maize silage no lower than 30 cm, leaving the potassium-rich, indigestible lower stems. The recommended stage for cutting maize for silage is full grain maturity (black layer at the base of the grain). Sunflower can also be used for silage.

Pasture and Grazing Management

The merits of continuous versus rotational grazing will not be discussed; in general, the former show superiority at lower stocking rates. Rotational grazing is generally preferred in ICLZT systems in tropical Brazil. On a pasture which is still in the maintenance phase, subdivision and rotational grazing on a 30-40 day interval can give a significant boost to production. In this situation, the investment in more animals to profit from the higher carrying capacity is greater than the investment in electric fences (now common) and watering facilities. Continuously grazed permanent pasture, or temporary winter pasture, both need to recover after the first rains; the latter to regenerate biomass for desiccation in order to provide adequate residue cover (at least 30 days). Temporary grazing on the crop area must allow grazing on part of the permanent pasture to be deferred until the onset of the rains, preferably in bottomland with better moisture availability. At this critical period beef prices are still high and stockyard or supplemental feeding with silage or green chop may be attractive alternatives.

Legumes in Pastures

The establishment and management of grass-legume pastures have, so far, been beyond the management capacity of most farmers in tropical Brazil, although adapted cultivars and experimental results are available. Where it has been done, results are highly encouraging, but management for legume survival is critical. Undersowing, oversowing, sowing the legume before the grass and seed pelleting or coating require more study.

Pigeon pea is a notable exception and its use in ZT sowing into pasture, is well-established and increasing. The most successful perennial legume has been *Stylosanthes guianensis* cv. Mineirão, but it is slow to establish and its seed is high-priced.

Liveweight gains on different pastures were maximized by using a *Stylosanthes guianensis* protein bank, with *Brachiaria brizantha,* giving gains of 577 kg/ha/year as compared to 258 kg/ha/year with *B. brizantha* alone. Another grass/legume pasture gave an intermediate performance. *Leucaena leucocephala* protein banks suffer from severe dry season leaf drop, making them only suitable to very restricted areas with a relatively high water table, but which do not become waterlogged in the rains. Protein banks avoid the complications of managing a mixed sward, and access once a week can be adequate.

6

Green Manure/Cover Crops in Conservation Agriculture

Soil degradation is the principal cause of a continuous decrease in crop production. The consequences of this are reduced economic income and increased poverty among rural families. One of the principal reasons for this fact is the continuous utilization of inadequate methods of soil management, including the burning of vegetative residues, excessive tillage, and monoculture. The exposure of bare soil to climatic agents accelerates the soil degradation process, as they cause excessively rapid decomposition of biomass and favor the erosion and leaching of nutrients.

The decrease in productivity is closely tied to a decline in the levels of soil organic matter. In poor soils, it is organic matter that determines the improvement of physical aspects, water retention, and biological activity, as well as the storage and slow release of nutrients. These aspects are most significant in sandy soils. Therefore, in order to maintain soil productivity in agricultural systems on small farms biomass is shown to be an essential element, due to the fact that it permits nutrient recycling and controls the microbial population that maintain favorable soil properties. One great difficulty in relation to the maintenance of biomass in tropical and subtropical climatic regions is that it breaks down much more rapidly than the capacity of conventional agricultural systems to replace it.

The strategies used by technicians and farmers to counteract the reduced yields caused by soil degradation are often based on the use of large quantities of inputs (fertilizers and pesticides), which results in high production costs. The apparent absence of agricultural systems for sustained productivity within the scope of small farmers is often due, not to lack of technology or to the low yield potential of traditional varieties. Rather, this situation is due to limited knowledge, or lack of awareness, on the part of technicians and farmers about practices that function in harmony with the environment in tropical and subtropical regions. At times, the absence of operative conditions (availability of credit, seeds, machinery, etc.) is a limiting factor in the development of this process.

Management measures to maintain soil fertility on small farms in Paraguay should be oriented toward the utilization of practices that maximize biomass production while minimizing its decomposition. In this sense, crop rotation, together with the use of green manure/cover crops and No-Tillage, form part of a technological strategy that has been proven by research and farmer practice to be efficient and economically viable, and that has as its objective the increase and conservation of soil organic matter. These practices together are referred to as "Conservation Agriculture".

Green Manure/Cover Crops

Green manure/cover crops (GMCCs) are plants that are grown in order to provide soil cover and to improve the physical, chemical, and biological characteristics of soil. GMCCs may be sown independently or in association with cro

In general, green manure/cover crops are used to pursue the following objectives:

— Provide soil cover for No-Tillage (reduces water evaporation and soil temperature, and increases water infiltration).
— Protect soil from erosion.
— Reduce weed infestation.
— Add biomass to soil (in order to accumulate soil organic matter, add and recycle nutrients, feed soil life).
— Improve soil structure.
— Promote biological soil preparation.
— Reduce pest and disease infestation.
— By carrying out these functions, green manure/ cover crops offer the following benefits:
— Increase economic return (when adequately chosen).
— Reduce need to use herbicides and pesticides.
— Increase yield and improve quality of the following crops.
— Prevent soil erosion.
— Conserve soil humidity.
— Maintain or increase soil organic matter content.
— Provide nitrogen to the soil.
— Improve soil fertility.
— Reduce fertilization costs.

Importance of Green Manure/Cover Crops

Soil Cover

One of the principal benefits of green manure/cover crops is the improvement of soil cover (live or dead). This has favorable effects on the physical, chemical, and biological properties of soil.

In conventional systems (tillage with plow, burning), in which bare soils predominate, the direct impact of raindrops breaks soil aggregates in the surface layer, resulting in the obstruction of the soil's pores. This causes the sealing of the soil surface and impedes the infiltration of water, which then runs off the surface, carrying with it part of the soil and producing what is known as water erosion.

In systems that promote soil protection, the principal effect of cover is to reduce erosion, either by preventing the direct impact of raindrops on soil and/or by reducing the velocity of surface runoff.

Cover also helps maintain soil humidity through shading and control of the thermal regime (lowers temperature, therefore reduces evaporation of water from the soil and increases water infiltration). These conditions also favor greater biological activity of the soil's flora and fauna (improves nodulation of legumes, increases the quantity of earthworms, etc.).

Plant cover, whether dead or alive, creates a microclimate that reduces the speed at which organic matter decomposes (mineralizes), favoring its accumulation in the soil. Furthermore, cover reduces the loss of organic matter from erosion. The degree of protection that a GMCC's live cover can offer the soil will depend principally on its initial speed of growth, growth habit (upright, creeping, etc.), the degree or volume of growth, as well as the duration of the plant's cycle.

Once flattened, or once the GMCC's cycle has ended, the durability of the dead cover over soil depends on the quantity and distribution of biomass produced. Likewise, it depends on the speed at which the organic matter decomposes, which, in turn, depends on the Carbon/Nitrogen ratio and on the lignin content of the residues (which varies by species). It also depends on environmental conditions that affect microbial activity (temperature, humidity, soil oxygenation, pH, etc.).

Maintenance and/or Accumulation of Organic Matter

Organic matter is the basis for soil fertility in agricultural systems because it performs multiple physical, chemical, and biological functions, and is the principal source of nutrients for plants through recycling. For these reasons, and because chemical fertilizers are not used on small farms, soil organic matter takes on a special importance.

Crop residues in conventional systems are not enough to compensate for the loss of organic matter, due to the high rate of mineralization that occurs in tropical and subtropical climates. On small farms, the only practical and economic means to maintain and/or increase soil organic matter levels is to utilize a green manure/cover crop, or a combination of GMCCs, that has a high potential for biomass production in order to complement the contribution of crop residues. The different species contribute different quantities of organic carbon.

One great advantage that the recycling of nutrients by green manure/cover crops offers is that they are released slowly to the soil as the dead residues decompose, allowing the plants that follow to make use of them gradually and more efficiently.

Weed Control

One of the aspects that farmers most appreciate about the use of green manure/ cover crops (GMCCs) is weed suppression. This allows them to save labor used for hoeing and to reduce the use of herbicides, lowering production costs. On small farms, where the possibility of using herbicides is limited due to lack of knowledge about their application and lack of access, GMCCs are essential for the success of Conservation Agriculture.

Work done at the Choré Experimental Station showed that weed infestation was markedly reduced in plots of cotton sown after summer green manure/cover crops associated with corn, in comparison with the plots where GMCCs were not used. According to these results, the GMCCs that displayed the best weed control were grey-seeded mucuna and sunnhemp, with reductions in infestation of 95 and 93%, respectively, in relation to the check plot without GMCC that had 2,676 weed plants/m^2.

The lower population of weeds in the cotton field due to the use of green manure/cover crops, besides favoring the crop by reducing competition for water, light, and nutrients, may represent an additional economic advantage for the farmer by saving the labor (hoeing) for weed control, which often leads to abandonment of the crop.Winter green manure/cover crops also reduced the incidence of weeds in the following corn crop.

In the Department of Paraguarí, in the cultivation of No-Till sugar cane, the inclusion of grey-seeded mucuna as cover for the first year, and of dwarf mucuna the second year, reduced the number of hoeings per year to one; the conventional system had to be hoed three times per year. Similar results were obtained in the Department of Caaguazú, with the use of grey-seeded mucuna as cover for the cultivation of watermelon.

Green manure/cover crops, because of their ability to suppress weeds, allow the adoption of the Conservation Agriculture on small farms without the use of herbicides.

Other experiences in the Department of Paraguarí showed that pigeon pea sown at high densities reduced the incidence of difficult-to-control weeds such as nutgrass

(Cyperus rotundus); and succeeded in eliminating grasses such as Jamaican crabgrass (Digitaria horizontalis). Also, grey-seeded mucuna reduced the infestation of sandbur (Cenchrus equinatus) on several farms in the Department of San Pedro.

Increase and Maintenance of Crop Productivity

When adequately chosen, green manure/cover crops (GMCCs) result in significant increases in the production of the crops that follow. However, in the long term, of interest is not only the increase but also the maintenance of high production. The use of GMCCs is indispensable for the achievement of this objective. The correct selection of green manure/cover crop species to precede a certain crop makes it possible to achieve a nigher yield and greater economic return.

Another important aspect of green manure/cover crops is that they make No-Tillage viable through biological tillage and the reduction of weed incidence. And so, with this system, the work of soil preparation (plowing, etc.) is avoided, and crop management practices (weed control, hilling-up, etc.) are reduced. This, added to the possibility of utilizing less fertilizers and other inputs (eventually herbicides and other pesticides), permits a reduction in production costs for those crops that follow GMCCs.

One more advantage of green manure/cover crops is that they require little investment of capital in that, normally, the farmer can produce and sow the seed himself.

Requirements that Should be Met

In order to achieve the greatest benefits from green manure/ cover crops (GMCCs), it's necessary to know all aspects related to them (family to which belong, growth habit, cycle, rusticity, weed competition, effects on soil, nutrient recycling, nitrogen-fixing or not, seed production, performance when faced by pests and diseases, ways to manage, etc.). Also, it's necessary to know the objectives being sought by including them, as well as aspects concerning the production systems in which they will be included (climate, soil type, conditions of soil fertility, crops with which they will be integrated in the system, available machinery and/or implements, etc.).

As a general rule, the species of green manure/cover crop (GMCC) selected should be rustic and require few crop management practices. Normally, when soils are apt for traditional agricultural crops (cotton or corn) it should not be necessary to fertilize GMCCs, nor to apply lime. An exception would be in those cases where a system fertilization strategy is used, placing fertilizer with the GMCCs and not with the commercial crop. However, when soils are very degraded, the utilization of fertilizers and lime should be considered in order to achieve good growth of the green manure/cover crops and so produce enough biomass to initiate recuperation of those soils.

There exist some requirements that green manure/cover crops should meet, that would make them more favorable for incorporation into an agricultural production system, the most important being:

— Have low establishment and management costs.
— Be easy to sow and manage.
— Achieve good shading and weed suppression.
— Produce a favorable residual effect on cash and subsistence crops.
— Be rustic and require few crop management practices.
— Present good conservation characteristics.
— Avoid proliferation of pests and diseases.

— Avoid competition for land, labor, time, and space with cash or subsistence crops.

The principal conservation characteristics that green manure/ cover crops should have and that should be considered in their selection are the following :

— Rapid growth and good soil cover under prevailing soil and climatic conditions.

— Production of a great quantity of green and dry mass, of the above-arts and roots.

— Slow decomposition of dry matter produced.

Most Adapted Species of Green Manure/Cover Crops

Green manure/cover crops (GMCCs) are differentiated by their period of growth: summer GMCCs grow during the spring and summer seasons, and winter GMCCs develop during the fall and winter seasons.

Spring-Summer Species

Grey-seeded mucuna (Mucuna pruriens = Stizolobium cinereum)

Characteristics: It is an annual, herbaceous, creeping, climbing legume. It is of medium size, has rapid initial growth, is rustic, tolerant of pests and diseases, and also controls nematodes. It produces a good quantity of dry matter under conditions of medium and high fertility (8 to 10 t/ ha); however it does not develop well on extremely degraded soils (2 to 4 t/ha). It possesses a vigorous root system capable of biological fixation of atmospheric nitrogen. It is an excellent green manure/ cover crop for the majority of crops that follow it in rotation, and the effect it has on corn, cotton, cassava, tobacco, and vegetables is outstanding. It also has a very marked effect on weed suppression, both during growth by smothering them and as dead cover through shading and allelopathy1.

Ways to use: One of the alternatives most practiced is the sowing of mucuna in association with corn. It is recommended that corn be sown early (August-September), in order to install the mucuna as soon as possible (November-December).The earlier it is sown during this period, the greater the production of biomass. For this system, it is recommended that 2 rows of mucuna be sown between every two rows of corn, 90 to 110 days after the corn was sown, and not before, to prevent the mucuna from smothering it.

Recommended spacing is 50 cm between rows and 40 cm between holes, with 1 to 3 seeds per hole. Seed use is approximately 100 to 120 kg/ha. The weight of 1,000 seeds varies from 1,000 to 1,300 gm, depending on the year.

Approximately 40 days after sowing the mucuna (depending on the corn variety), the corn is harvested manually or left among the mucuna until winter. Grain conservation is good in the latter case, with less attack by insects, especially weevils. After the corn is harvested, the mucuna is left to grow until it is time to flatten it. It is recommended that mucuna be flattened with a knife-roller 15 days before sowing the following crop, in order to crush the cover but without giving time for new weeds to grow or for the biomass to degrade. When there are frosts, or the mucuna dies because its cycle has ended, it's generally not necessary to flatten the mucuna. Otherwise, it may be flattened in July-August with a knife-roller, machete, or herbicides. If there are viable seeds, it will be necessary to harvest them in order to prevent them from resowing naturally and becoming weeds in the following crop.

Another option to use grey-seeded mucuna is to sow it following summer crops that are harvested in January-February (corn, peanuts, tobacco). The field often have few weeds and so there is no need for weed control prior to sowing the mucuna. If necessary, weeds can be controlled with a knife-roller, by hoeing or cutting, or with herbicides. The sowing method and density are the same as recommended for the association of mucuna with corn.

Grey-seeded mucuna is one of the species most utilized in production systems on small farms with soils of medium to high fertility, principally in association with corn.

Between the end of mucuna's cycle and the sowing of crops such as cotton and sesame (October-November), there is a period of about 2 months with no crop. In order to avoid the proliferation of weeds during this period and to obtain a greater production of biomass and cover, it's possible to advance the flattening of mucuna to June and sow a winter green manure/cover crop, preferably a mixture of black oats, white lupine, and oilseed radish.

The best time to sow grey-seeded mucuna is in November. Sowings of January and later generally produce little biomass.

Seed production: Normally, seed can be harvested from mucuna used as a green manure/cover crop if it was sown early. However, for greater seed production it is recommended that mucuna be sown at a lower density (12 to 15 kg/ha), with support such as bamboo tripods or tree trunks and branches, or in association with agricultural crops such as corn, old cassava, castor, etc. In this way the rot of flowers and seedpods from contact with the soil is avoided. Fences may also be used as support.

One practical system is to use corn as a support, sowing the mucuna around 30 days after planting the corn. In order to avoid possible losses due to frost, mucuna should be sown in October. In this system it is recommended that 1 row of mucuna be sown between every other row of corn, leaving 2m between rows and 90 cm between holes, and depositing 2 seeds per hole (12 to 15 kg/ha of seed). Mucuna's cycle ends in July-August at 240 to 270 days after sowing. Expected yield of seed is from 1,000 to 1,500 kg/ha. It is important to classify seeds, discarding those that are rotten or with black coloration (they won't germinate). A seed-producing area of 1,000 m^2 is needed to produce enough seed to sow 1 ha of green manure/cover crop.

Pigeon pea (Cajanus cajan L. Millsp.)

Characteristics: It is a semi-perennial (2 to 4 years), shrubby, tall legume. It is characterized as being very rustic in terms of fertility and soil type. It tolerates both drought and cold. When it freezes, the leaves dry out but it resprouts immediately. In the case of a hard frost, part or all of the plant may die. It produces great volumes of biomass annually, even on extremely degraded soils (7 to 14 t/ha); for this reason, it is an excellent option to initiate soil recuperation under these conditions. Initial growth is slow, therefore weed infestation may occur. In this case, it is necessary to hoe in order to achieve good initial growth. Later, it covers the soil well and controls weeds with its shade. Besides the species mentioned there exists a dwarf pigeon pea, which is quite a bit shorter and has smaller pods and seeds.

Ways to use: It is recommended principally for the recuperation of degraded soils, grown in association with corn: a) as an alternative to traditional fallow, leaving the plants to grow for 2 to 4 years; or b) as the first GMCC in an annual crop rotation system. These alternatives may also be practiced on more fertile soils. The most appropriate moment to sow pigeon pea is 60 to 70 days after sowing corn, normally after the second hoeing, but it may be moved forward to 30 days to coincide with the first hoeing. It is recommended that 2 rows of pigeon pea be sown between each two rows of corn, placing 4 to 6 seeds per hole at a distance of 30 cm between holes and using 25 to 35 kg/ha of seed. The weight of 1,000 seeds is approximately 140 to

When used in rotation with annual crops, it is recommended that pigeon pea be flattened 2 to 3 weeks before the following crop is sown. This operation may be done with a knife-roller, later cutting with a machete any stems that rise above the soil level so that they don't resprout. This method reduces by half the time used when flattened with only a machete.

Pigeon pea, because of its rusticity and high production of biomass, is an ideal plant to recuperate extremely degraded soils.

Seed production: It is possible to take advantage of the same sowings intended for green manure/cover crop, leaving a strip not flattened. Better seed production is achieved in crops that are less dense; to that effect, it is possible to sow 1 or 2 rows of plants in the form of live fences. The seedpods of the majority of tall varieties mature in July-August, approximately 210 to 270 days after sowing, while those of the dwarf varieties mature in May-June (180 to 210 days after sowing). Seeds may be harvested by collecting the dry pods manually, or by cutting the plants; they are later dried and threshed. Seed yields generally vary from 1,000 to 2,000 kg/ha.

The seeds are highly sensitive to weevils, and attack frequently occurs while the pods are still on the plant. It is recommended that seeds be treated with ash, sand, or lime, mixing them in a container. Also, the seeds may be treated with aluminum phosphide (Phostoxin or Gastoxin) immediately after harvest if they are weeviled, and the lot controlled periodically to detect later attack. A seed-producing area of 300 m^2 is needed to produce enough seed to sow 1 ha of green manure/cover crop.

Jack bean (Canavalia ensiformis L. DC)

Characteristics: It is an annual, herbaceous, erect legume that climbs little, is of medium size, and grows vigorously.

It is characterized as being less demanding in terms of soil fertility than grey-seeded mucuna. It adapts very well in all of Paraguay's Eastern Region, even withstanding frosts of low intensity. It is also tolerant to drought. It normally has no diseases and, although occasionally attacked by leaf-chewing insects, succeeds in achieving good cover and producing important quantities of biomass in soils of medium and high fertility (approximately 6 to 7 t/ha of dry matter). In extremely degraded soils it produces 2 to 4 t/ha of dry matter.

Jack bean attains rapid initial soil cover due to the great size of its leaves. It competes very well with weeds, an effect that lasts even when cover is reduced by leaf fall. It possesses a vigorous taproot with an outstanding capacity to decompact soil (biological tillage). In spite of early production of the first seedpods, its vegetative development continues for a long period of time (10 to 12 months) during which it again produces seed.

Ways to use: Due to its growth habit of climbing little, it can be cultivated simultaneously with other annual crops (1 row between each two rows of crop), principally with corn or with cassava when sown beginning in August. This is an advantage because soil cover is present practically all year, with good weed suppression and nitrogen added by biological fixation. However, in dry years simultaneous cropping has caused a reduction in corn yields in Choré, due to competition for water. Another advantage is that, because of its long cycle and cold tolerance, crops such as cotton and sesame may be sown without a break, immediately after passing the knife-roller.

Jack bean may also be sown after the crop's first or second hoeing (30 or 60 to 70 days), or later. In perennial plantations (citrus, yerba mate, etc.) it is very efficient as a green cover crop because it doesn't send out climbing branches and it covers the soil for a prolonged time. In sowings that are simultaneous, or at 30 days, it is recommended that one row be sown between every two crop rows, with holes at 30 cm and 1 or 2 seeds per hole. In this case 70 to 80 kg/ha of seed are used. In later sowings (as of 60 days), it is recommended that 2 rows be sown between every two crop rows, which raises seed use to 140-160 kg/ha. The weight of 1,000 seeds ranges from 1,300 to 1,500 g. It may be flattened with a machete or knife-roller, immediately or 10 to 15 days before the sowing of summer crops (cotton, corn, etc.). Any weed infestation that occurs later should be adequately controlled.

Seed production: Seeds may be produced in the same field used for a green manure/cover crop. They may also be produced in a pure crop, using a spacing of 1 m between rows and 30 cm between holes. The seeds mature erratically, and should be harvested as they ripen. This occurs from approximately 4 months after sowing until around 7 months after (March-July). Later, the seedpods need to be dried in the sun to facilitate threshing. Seed yields range from 1,000 to 1,500 kg/ha. A seed-producing area of 750 m^2 is needed to harvest enough seed to sow 1 ha of green manure/cover crop.

Dwarf Mucuna

Characteristics: It is an annual, herbaceous, medium to short legume, with an erect, determinant, and non-climbing growth habit. Production of dry matter is around 4 t/ha in soils of medium and high fertility. It develops little biomass in extremely degraded soils (2 to 3 t/ha of dry matter). It has good capacity for nitrogen fixation and also promotes the control of soil nematodes. In general, there are no pests or diseases that severely affect its development. Dwarf mucuna is affected by frost, which can cause it to die when intense. It achieves good weed suppression in spite of its short cycle and determinant growth. It provides less soil cover during its development than other species of green manure/cover crops.

Ways to use: Dwarf mucuna can be associated with annual summer crops (corn, cassava, etc.) and also with perennial or semi-perennial crops (yerba mate, sugar cane, citrus, pineapple, and others) to precede winter crops. When associated with annual summer crops, it is recommended that 1 row of dwarf mucuna be sown between every two crop rows, with a manual jab planter or sharpened stick. It may be sown simultaneously or up to approximately 60 days after the corn is sown, in holes 40 cm apart and with 2 to 3 seeds per hole. In simultaneous sowings during dry years there may be competition for water. When associated with perennial or semi-perennial crops it may be sown in October-December with a spacing between holes of 50 x 40 cm, with 2 to 3 seeds per hole (70 to 90 kg/ha). The weight of 1,000 seeds varies from 530 to 750 grams. In general, it is not necessary to flatten dwarf mucuna since its cycle ends in April, which allows a winter crop to be sown directly over its residues.

Seed production: It is possible to harvest seed from the green manure/cover crop. Dwarf mucuna's cycle, from sowing to harvest, is 150 to 180 days. It's possible to reach seed yields of 800 to 1,500 kg/ha. Dwarf mucuna's seedpods rot easily when they remain in contact with the soil and there is rain, and so there may be an important loss of seeds if they are not harvested as soon as they ripen (March-May). A seed-producing area of 500 m^2 is needed to harvest enough seed to sow 1 ha of green manure/cover crop.

Sunnhemp (*Crotalaria juncea* L.)

Characteristics: It is an annual, tall (over 3m) legume, that is characterized as having great biomass production (7 to 8 t/ha of dry matter when sown in January in association with corn). It also grows well on degraded soils and develops much better than mucuna. Its initial growth is rapid and it has an excellent suppressing effect on weeds. It also stands out for its ability to lower nematode populations.

Ways to use: Sunnhemp has an excellent residual effect on the majority of crops, both summer and winter (vegetables and others). It adapts well to late sowings (January-February) after the harvest of corn, peanuts, cowpeas, etc. It is also advisable to associate it with corn as an alternative to mucuna, sowing it 60 to 70 days after the crop was sown.

When associated with corn, sunnhemp may be sown broadcast, incorporating seed into the soil with a hoe (when possible, taking advantage of a weeding). Also, it may be sown in furrows (47 to 48 seeds/m) or in holes (30 cm between rows with 14 seeds per hole) with 2 rows between each two rows of crop. If not associated with a crop, sunnhemp may be sown in the same way, using a knife-roller or a disc harrow (regulated to roll over the ground, cutting as little as possible into the soil) to incorporate the seed when sown broadcast. Sunnhemp is also very appropriate for sowing in the season prior to the establishment and renovation of sugar cane (reduces nematode populations and adds nitrogen). For all systems proposed it is recommended that 40 kg/ha of seed be used. The weight of 1,000 seeds is around 40 to 45 g.

It may be flattened with a knife-roller or machete, immediately before the following crop, preferably leaving no more than a week between flattening of the sunnhemp and the sowing of the crop. This management is most suitable for crops to be planted in August-September (corn, tobacco, melon, watermelon, cassava, etc.), because the sunnhemp is found in its optimum moment to be flattened (flowering – seedpods starting to fill). For crops that are sown later (cotton, etc.), it is recommended that flattening be delayed until just before the plants have viable seed. Sunnhemp may also be left until its cycle ends, as long as seedpods are harvested since the seeds produced could become weeds.

Seed production: Fields sown for green manure/cover crops normally produce a good quantity of seed. If present, carpenter and bumble bees (Xylocopa sp., Bombus sp.) pollinate the flowers and assure good yields. If the objective is specifically to produce seed, it is advisable to sow long, narrow strips to favor the activity of the bumble bees. There generally occur attacks of the Bella Moth (Utetheisa ornatrix) on seedpods, impairing seed formation and ripening. However, it doesn't cause great losses under the conditions of Paraguay's Eastern Region. Seedpods ripen in June-July, around 180 to 240 days after sowing. It may be harvested manually, cutting branches with seedpods, or the pods themselves, and threshing them later with a stick. Yields vary around 600 to 800 kg/ha, and may reach 1,200 kg/ha or more. A seed-producing area of 700 m^2 is needed to harvest enough seed to sow 1 ha of green manure/cover crop.

Pearl Millet

Characteristics: It is an annual, erect, tall grass, able to reach a height of around 2 to 3 m. In soils with medium and high fertility it has great potential for biomass production (over 10 t/ha of dry matter), which allows it to contribute a great quantity of organic matter and recycle large quantities of nutrients, principally potassium. In conditions of low fertility its potential for biomass production falls drastically (around 2 to 3 t/ha of dry matter) and in extremely degraded soils it grows very little.

It grows rapidly and competes very well with weeds. In addition, its great quantity of residue, which decomposes slowly after being flattened, leaves an excellent cover for a long time. It has high drought tolerance, developing in regions with precipitations that start at 200 mm. It can withstand soils that are acidic and saline. It has a high potential to tiller and resprout, and serves as forage for animals.

Ways to use: It is recommended to precede crops such as so , cotton, cowpeas, etc. as short-period cover, due to its rapid initial growth and short cycle, principally in the warmer regions of Eastern Paraguay (Central and North). In this case it is sown in September-October. In addition, it is suitable for late sowings (January-February), after the harvest of crops such as peanuts, cowpeas, corn, etc. It may be used for forage (grazed or cut and carry), in which case it should be left to resprout in order to produce enough biomass to cover the soil until the following crop is sown. If pearl millet's cycle ends (whether used as forage or not) the seeds may be harvested; otherwise, they will probably be eaten by birds.

Twenty to 25 kg/ha of seed are used in broadcast sowings, and the seed incorporated into the soil surface with a knife-roller, disc harrow (regulated to roll over the ground, cutting as little as possible into the soil), etc. The weight of 1,000 seeds of the common varieties is around 4 to 10 g. It may be flattened with a knife-roller 2 to 3 weeks before the following crop is sown, complemented by herbicides, hoe, etc. in order to control eventual resprouting. It is recommended that pearl millet be associated with a legume such as pigeon pea, sunnhemp, cowpea, or other.

Seed production: Seeds may be harvested from the fields of green manure/ cover crop. In order to achieve good yields, it is essential to take note of the risk that seed may be eaten by birds. It is recommended that seeds be harvested as soon as they ripen (May-June) and dried immediately. The harvest may be manual. Heads with dry seeds may be placed in bags (preferably cloth), rubbed by hand to remove the grains, and later winnowed. Pearl millet's complete cycle is from 120 to 150 days and seed yield may reach about 1,000 kg/ha. A seed-producing area of 250 m^2 is needed to harvest enough seed to sow 1 ha of green manure/ cover crop.

Forage sorghum (Sorghum sp.)

Characteristics: It is an annual, erect, tall grass. It has rapid initial growth, and covers and protects the soil, controlling weeds well. In addition, it has an allelopathic effect on weeds. It can produce great quantities of vegetative mass in barely two months of growth. Production of dry matter varies from 5 to 10 t/ha depending on variety used, time of sowing, time of development, and soil fertility. Under good conditions it can produce around 20 t/ha of dry matter. It performed better than pearl millet on very degraded soils in the Department of Paraguarí. In addition to forage sorghum, it is possible to utilize grain sorghum; this also develops well, produces great quantities of mass, and makes it possible to achieve a good grain harvest.

Ways to use: It is used principally as a short-period green manure/cover crop in sowings as of August-September, being able to develop for 50 to 70 days until crops such as cotton, sesame and soybeans are sown. It can also be used in late sowings of January-February after the harvest of corn, cowpeas, peanuts, etc., and left until the next summer crop is sown. In this situation, the seed is normally harvested, and the residues then cut flush with the soil in order to take advantage of it as forage and to favor resprouting. With this a good quantity of biomass and cover can be produced. Also, it may be grazed directly by animals and then left to resprout, taking care that it is not grazed before 35 to 45 days after resprouting.

It is recommended that forage sorghum be sown broadcast using 20 to 25 kg/ha of seed, incorporating them superficially into the soil with a knife-roller or disc harrow (regulated to roll over the ground, cutting as little as possible into the soil). The weight of 1,000 seeds of the variety Sudan Grass is around 15 g. It may also be sown in mixture with other species. Management to flatten it is done 2 to 3 weeks before the following crop is sown, with a knife-roller; this may be complemented later with an herbicide, hoe, etc. to eliminate sprouts.

Seed production: Seed may be harvested from fields of green manure/cover crops, regardless of the system used. The harvest may be done by cutting the heads with a knife, sickle or pruning shears, and threshing them later. Seeds mature in May-June, at 120-150 days after sowing. Yields generally surpass 1,000 k/ha of seed. A seed-producing area of 250 m^2 is needed to harvest enough seed to sow 1 ha of green manure/cover crop.

Lablab (Lablab purpureus L., or Dolichos lablab L.)

Characteristics: It is a biennial, herbaceous legume of medium size, which has a creeping and climbing growth habit similar to that of grey-seeded mucuna, though it is not as aggressive nor does it produce as much biomass (4 to 6 t/ ha of dry matter). It is outstanding for its good growth under drought conditions. It is susceptible to nematodes and may increase their populations in the soil. Its long cycle provides cover for an extended time if there are no hard frosts, being able to perform as a biennial plant. Its forage is of better quality than mucuna's and, above all, it is very uitable for making silage in mixture with grasses. On occasion it may suffer attacks of pests such as the beetle of the South American rootworm (Diabrotica speciosa), bean leaf beetle (Cerotoma sp.), and aphids. There are varieties with different seed colors (black, brown, and white) that have different lengths of cycle (short, medium, and long, respectively).

Ways to use: It can be used in the same system recommended for grey-seeded mucuna (associated with corn), principally in places that are dry and of lower fertility, as long as there are no nematode problems. One advantage it presents is that, being less frost sensitive, it is possible to have live cover up until the moment it is flattened in order to sow late crops (cotton, sesame, and others).

It is advisable to sow lablab starting from 60 days after the corn was sown, placing 2 rows of lablab between each two rows of crop , leaving 30 cm between holes and 3 to 4 seeds per hole (about 40 to 50 kg/ha). The approximate weight of 1,000 seeds is 260 g for brown-seeded varieties, 250 g for those with black seeds and 230 g for those with white seeds. Management to flatten it may be done 10 to 15 days before the following crop, with a knife-roller or machete, eliminating new sprouts in the same way or with herbicide.

Seed production: Seed may be harvested from the green manure/cover crop. The seedpods of lablab are ready for harvest around July-August, depending less on the date sown and more on the variety (white-seeded varieties mature about 20 days later than those with brown seeds and these, in turn, approximately 20 days later than those with black seeds). This means greater difficulty in producing seed of varieties with lighter seeds in years and places with greater frequency of frost. Lablab has serious storage problems due to weeviling. In addition, it loses its ability to germinate the following year under normal conditions of conservation utilized on small farms, and for that reason seed should be produced each year. Seed yields vary from 500 to 1,000 kg/ha. A seed-producing area of 900 m^2 is needed to harvest enough seed to sow 1 ha of green manure/cover crop.

Forage peanut (*Arachis pintoi* L.)

Characteristics: It is a perennial, herbaceous, tropical and subtropical legume, creeping but not climbing, and of low growth habit. It displays excellent development on sandy soils, and also on clay soils of medium fertility. Its creeping growth habit is initially slow, but it covers the soil very well once established. It has good biomass production in soils of medium to high fertility (up to about 8 t/ha/year of dry matter) and produces almost no biomass in extremely degraded soils (less than 1 t/ha of dry matter). It generally has no problems with pests or diseases. In addition, it helps to reduce nematodes in the soil. Its sanitary performance is superior to that of grassnut (Arachis prostrata), which may suffer attack by thrips and nematodes.

Ways to Use: It is normally suggested as a cover crop between rows of perennial crops such as yerba mate , fruit trees, etc. It is recommended that forage peanut be broadcast sown in October-November, using 8 to 15 kg/ha of seed, and incorporated with an animal drawn disc harrow (regulated to roll over the ground, cutting as little as possible into the soil) or knife-roller. The weight of 1,000 seeds is 161 g. It may also be multiplied by cuttings planted at a distance of 1 to 2 m from each other.

In systems with perennial crops, forage peanut may be left without flattening for several years. In rotation with annual crops it may be flattened with herbicides 10 to 15 days before sowing the following crop, or by opening furrows (with hoe or ripper), or by making narrow paths with a hoe, machete, herbicide, etc. where the crop will be sown. Farmers in Brazil manage peanut by drying it with low doses of herbicide to partially burn the plants, then sow No-Till corn. Afterwards, the peanut resprouts and again covers the soil. It is utilized later as forage or cover.

Seed production: It is possible to obtain seed from fields of green manure/ cover crops. It is harvested manually in April-May, completing a cycle of 180 to 240 days. Yield is approximately 500 kg/ha. Harvest is difficult, therefore the price of seed is high. It is recommended that small quantities of seed be purchased and then multiplied in the field. A seed-producing area of 250 m^2 is needed to harvest enough seed to sow 1 ha of green manure/cover crop.

Creeping indigo (Indigofera endecaphylla L.)

Characteristics: It is a perennial legume of indeterminate habit, herbaceous, creeping, and of medium height. It tolerates drought and light frost. When burned by frost, it resprouts when temperatures rise, again covering the soil. It displays good capacity for nodulation, excellent soil cover, and resows naturally. It is outstanding for its ability to compete against weeds, which it smothers. It has no insect or disease problems, however it is recommended that this crop not be sown in areas with nematodes. It grows in clay soils as well as in sandy, acidic soils of low fertility (principally lacking phosphorous). It can produce 5 to 6 t/ha of dry matter.

Ways to use: It adapts well when interplanted with perennial crops such as yerba mate, citrus, grapes, macadamia, mango, etc. In this case, it is recommended that it be sown in the months of October-November. When sown broadcast, 15 to 20 kg/ha of seed are used. The weight of 1,000 seeds is 6 g. Another option is to sow the creeping indigo in rows 50 to 70 cm apart (seeds sown by trickling). In addition, it may be propagated by cuttings. Once established it may be left to grow for several years. Creeping indigo may be managed by drying it with an herbicide, or opening furrows (with a hoe or ripper) through the live cover, or making a narrow path with a hoe, machete, herbicide, etc. to sow a crop.

Seed production: Seeds may be harvested from the fields of green manure/cover crop. They generally reach maturity in September-October, 9 to 12 months after sowing, and may be harvested each year. It is possible to produce 60 to 150 kg/ha or more of seed, which is harvested by hand. A seed-producing area of 3,000 m^2 is needed to harvest enough seed to sow 1 ha of green manure/cover crop.

Tephrosia (Tephrosia tunicata L., Tephrosia candida L.)

Characteristics: It is a perennial, tropical and subtropical legume; it is bushy, has woody stems, is 3 to 4 m tall, and shows slow initial growth. It stands out for its great rusticity, growing well in soils that are clay, acidic sandy, and of low fertility. Biomass production of Tephrosia candida varies from 7 to 15 t/ha/year of dry matter. It reduces nematode populations. It has a vigorous taproot, capable of decompacting soil. It may be used to control "vaquitas" (beetles of Diabrotica spp.). It has good residual effect, and increases the yields of cash or subsistence crops sown afterwards. It is highly resistant to attack by pests.

Ways to use: Tephrosia can be utilized in several production systems: a) For the recuperation of extremely degraded soils, since it can be left for 2 to 5 years, according to the degree of soil degradation. Tephrosia may be sown in rows every 60 cm to 1 m, with 10 to 12 plants per linear meter, or with a manual jab planter with 30 cm between holes and 3-4 seeds per hole. Eight to 15 kg/ha of seed are used. The weight of 1,000 seeds is 15 g for Tephrosia candida and 28 g for T. tunicata. After the soil has been recuperated, tephrosia can be flattened by cutting the plants flush with the soil and eliminating any sprouts that appear. Because of their woody stems, they are tough to flatten with a machete.

As a shrub, it may be used in live fences/vegetative barriers, or in alley cropping. Rows should be separated by 4 to 12 m, principally in areas with steep slopes. Later, it may be cut (pruned) with a machete and the branches and leaves distributed over the soil; afterwards, a crop such as corn, cowpeas, cassava, cotton, peanuts, etc., may be sown. The resprouts are also cut (1 to 2 cuts) and placed between the rows of growing crops. As a cover in association with perennial crops. It may be cut and left between the rows of crops such as yerba mate and others.

Seed production: The cycle is complete in August-September, 240 to 330 days after sowing. The harvest may take place annually during the same period of time. The harvest and threshing are done manually; seed production is from 200 to 400 kg/ha. On average, a seed-producing area of 600 m^2 (according to minimum yield) is needed to harvest enough seed to sow 1 ha of green manure/cover crop.

Leucaena (Leucaena leucocephala L. de Wit)

Characteristics: It is a tree of the legume family, perennial, and of tropical and subtropical climates. When the plants are young (less than 1 year) they cannot withstand frost. Adult plants are affected little by frost; part of the leaves and branches may be burned, but later resprout. It adapts well to a wide range of soils, but doesn't grow well in very acid soil. It is characterized by its great potential for nitrogen fixation and recycling, which may reach over 600 kg/ha/year with 3 to 4 cuts. Furthermore, it has a great capacity to recycle other nutrients, which previously leached (washed), through its deep roots. Like the majority of tree species, its early development is very slow. Its leaves and branches are of high nutritional value for animals, with a high protein content; however, there are limitations for its use as forage.

Ways to use: In agricultural production, leucaena is recommended principally for the recuperation of extremely degraded soils. It is grown in strips of 1 to 2 rows, leaving 6 to 12 m between rows in which annual crops may be grown. With less space between plants (50 cm between holes or trickled into a furrow), a greater quantity of biomass is produced (10 to 12 t/ha of dry matter with rows 6m apart). The weight of 1,000 seeds is around 50 to 60 g. It may be sown in the spring using a manual jab planter (3 to 4 seeds per hole), or seedlings transplanted in autumn (50 cm between holes and 6 m between rows) if a seedbed had been made, or by taking advantage of seedlings from natural regeneration. In order to sow seeds, which are hard when not recently collected, it's necessary to treat them beforehand with hot water to facilitate germination. One option is to pour boiling water over the seeds and stir continually for 2 to 3 minutes. They are then rinsed with cold water to be sown later.

Due to its initial slow growth, for leucaena to establish well it's necessary to control any weeds that emerge after sowing or transplanting. In order to avoid hoeing exclusively for that reason, it is recommended that leucaena be associated with annual crops, principally cassava. In this way, advantage is taken of the crop's weeding while at the same time a suitable microclimate is created for the growth of leucaena. In this system, cassava and leucaena seed are planted simultaneously. In the case of seedlings, it's possible to transplant them later, in the spring.

Leucaena is managed by cutting it periodically (3 to 4 times a year, mainly in spring-summer), at a height of 20 to 50 cm from the soil, before it produces viable seeds that could become weeds. Its fine branches and leaves are placed over the soil surface between rows. As they break down, they are taken advantage of by the crops. The thicker branches may be used as firewood or stakes or, preferably, left along the strips of leucaena where they will decompose. The organic matter that accumulates this way is later redistributed between the rows.

Seed production: Since it is a perennial crop, it is recommended that only 1 or 2 trees be left as a seed bank. The objective would be to replace plants in case of loss, or if there is need to expand the area under the proposed system. These trees should be outside the farmed area in order to avoid weed infestation from natural regeneration. Seed matures in July-August or all year, depending on the variety. It may be harvested annually. It is possible to assure that you will obtain enough seed to sow 1 ha of green manure by leaving 2 or 3 trees for seed production.

Fall-winter species

Black Oats (Avena strigosa Schreb)

Characteristics: It is an erect, annual grass of medium height, and very well adapted to the soil and climatic conditions of Paraguay's Eastern Region. It is resistant to attack by rust and aphids, therefore doesn't require any special crop management practices. It produces 4 to 5 t/ha of dry matter in soils of medium and high fertility, but develops little biomass in very degraded soils (1 to 3 t/ha of dry matter). It responds notoriously well to chemical fertilization, principally nitrogenous. In poor soils, even when it produces little biomass, it controls weeds well (allelopathy). Because of its rusticity it surpasses white oats, yellow oats, and rye. There is a variety named IAPAR 61, which has a longer cycle and produces a greater quantity of biomass than the other black oats. It displays a high capacity to tiller, and is able to produce 15 to 35 tillers per plant.

One of the principal benefits of black oats is the excellent soil cover it provides after being flattened, which lasts longer than that of other green manure/cover crops. Shading by the dead cover, added to the very strong allelopathic effect, allows for a high degree of weed suppression; in some situations, it does away with the need for other weed control practices such as hoeing or the use of herbicides.

The use of black oats improves soil health and also promotes increased yields in leguminous crops such as soybeans, cowpeas and Phaseolus beans that follow it in rotation. It is also optimal forage, able to withstand direct grazing in winter and with a good capacity to resprout, which can be taken advantage of as cover.

Ways to use: It is recommended that black oats be sown manually, distributing the seed broadcast with good soil humidity. To favor good germination, it is recommended that an animal-drawn disc harrow (regulated to roll over the ground, cutting as little as possible into the soil), a knife-roller, or a "rake" made of branches be passed after the seed is distributed. The sowing density (broadcast) recommended for pure crops is from 60 to 80 kg/ha of seed for common varieties and 50 kg/ha for the variety IAPAR 61. The weight of 1,000 seeds is from approximately 14 to 20 grams, the variety IAPAR 61 having smaller seed. The recommended sowing time for best plant development is April. However, black oats may be sown from March until the end of winter, which allows it to be incorporated into different production systems. Late sowings run the risk of insufficient rainfall in July and August, and the production of vegetative mass is generally quite reduced.

Black oats may be managed with a knife-roller, flattening the plants 2 to 3 weeks before the following crop. This preferably occurs when the grains are milky, around 120 days for common varieties and 150 to 160 days for the variety IAPAR 61.

Black oats may resprout when flattened early, which would make it necessary to desiccate it with an herbicide, or open furrows (hoe or ripper), or make narrow paths (hoe, machete, herbicide, etc.). On the other hand, when flattened with mature seeds, these may germinate and become weeds in the following crop. It is recommended that black oats be associated with white lupine to provide more nitrogen.

Black oats are outstanding for the excellent soil cover they produce, which lasts longer than that of other green manure/cover crops.

Seed production: It's possible to produce seed in the same field destined for green manure/cover crop. Because of its rusticity, it is exempt from the need for any cultural practices or phytosanitary treatment. On poor soils, black oats need to be fertilized. Seeds mature in August-September, around 140 to 150 days after sowing for common varieties, and 170 to 180 days for the variety IAPAR 61 (September-October). This variety is demanding of cold in the winter in order to achieve good seed production. On small farms the harvest may be manual, either pulling seeds off by hand, or cutting the plant with a machete or sickle to be threshed later with a stick. Approximately 600 to 800 kg of seed may be harvested per hectare. A seed-producing area of 1,200 m^2 is needed to harvest enough seed to sow 1 ha of green manure/cover crop.

White lupine (Lupinus albus L.)

Characteristics: It is an annual, herbaceous, erect legume of medium height. It is an excellent nitrogen fixer by means of bacteria that form nodules on the roots, adding approximately 90 kg/ha of nitrogen. Furthermore, it has a deep tap root system (1 meter or more) that improves the soil's physical condition (through decompaction) and recycles great quantities of nutrients.

White lupine is best adapted to the northern and central regions of Eastern Paraguay, where temperatures are higher and rainfall is less, although it is able to withstand temperatures of -3ºC to -4ºC. During initial growth it is sensitive to drought, but once its root system is developed it grows well with temporary water deficits. In more humid regions, such as in the Departments of Itapúa and Alto Paraná, or when conditions of high precipitation occur after sowing in Eastern Paraguay's central region (Departments of Paraguarí and Cordillera), it frequently suffers attack by anthracnose, a disease that can kill the plant and limit its cultivation. This problem decreases with crop rotation and, above all, by sowing white lupine in mixture with gramineous species, principally black oats. In this way, it's also possible to compensate for loss of cover caused by the disease. Never harvest plants infected with anthracnose when the grain will be used for seed. Production of white lupine biomass varies little between soils of different fertility (around 4 t/ha of dry matter) when sown densely and with at least 80 kg/ha of seed.

Ways to use: It is appropriate to precede crops that are demanding of nitrogen, such as corn and cotton. It is recommended that white lupine be sown in April, with a manual jab planter and at a spacing of 50 to 70 cm between rows, 30 to 40 cm between holes, and 3 to 4 seeds per hole (60 to 80 kg/ha of seed). The weight of 1,000 seeds is from 350 to 400 g. A delay in sowing almost always results in an appreciable decrease in growth and biomass

White lupine, like other leguminous green manures, adds large quantities of nitrogen that improve the yields of later crops such as corn and cotton. production. When sowing white lupine, it is important to pay attention to seed depth; if seeds are deeper than 4 cm, emergence is difficult and they may not even germinate.

The scarce cover of white lupine during its initial phase of growth allows weeds to proliferate, especially when the distance between rows is wide. To counteract this problem, it is recommended that it be associated with other species such as black oats , or that sowing density be increased by decreasing the distances between rows and between holes. In some zones, it may be necessary to inoculate seeds with the specific bacteria (Rhizobium lupini) to assure or improve nitrogen fixation. It may be flattened with a knife-roller and/ or machete immediately before the following crop.

Seed production: Part of the green manure/cover crop can be left to mature for seed production, in which case it may be necessary to control weeds. It completes its cycle in

October-November, approximately 180 days after sowing. It can be harvested by cutting the plant with a machete, and later threshed with static threshers, by beating the dry seedpods with a stick, or simply shelled by hand. Seed production varies from 1,300 to 2,200 kg/ha; one harvest was reported of 3,000 kg/ha on a mechanized farm in the Río Verde Colony (Department of San Pedro). Seed storage under natural, cool conditions (18 to 25 ºC) for a year, or in a cold storage room, is the best way to reduce to almost zero the level of anthracnose inoculum (Cardoso & Telardi, 1990). An alternative for small farms could be to dig a hole and bury the seed, properly protected with plastic "canvas". A seed-producing area of 500 m^2 is needed to harvest enough seed to sow 1 ha of green manure/cover crop.

Oilseed radish (Raphanus sativus L. var. oleiferus Metzg)

Characteristics: It is an annual plant that belongs to the cruciferous family, is herbaceous, erect, and of medium height. Initial growth is very fast. It's not very rustic, but produces a great quantity of biomass in conditions of medium and high fertility (up to 5 t/ha of dry matter). Its use is not recommended on extremely degraded soils, since its development is very poor (less than 2 t/ha of dry matter).

Some varieties stand out as having a deep tap root, capable of breaking compacted layers of soil. It also has the capacity to recycle great quantities of nutrients (principally nitrogen) and to make soluble some elements (phosphorous) that cannot be used by plants, making them available for the following crops.

Oilseed radish suppresses weeds very efficiently because of its rapid shading of the soil and its allelopathic effect. The rapid degradation of its biomass right after it is flattened or the end of the plant's cycle, allows rapid weed growth and the occurrence of problems related to low coverage. This situation can be improved by associating oilseed radish with other species of which the biomass decomposes more slowly, for example black oats.

The vigorous roots of oilseed radish are capable of loosening the soil and of breaking up compacted layers (biological tillage), favoring development of the crops that follow.

Ways to Use: Oilseed radish shows good residual effect on corn and other crops such as tomato, Phaseolus beans, peppers, etc. However, it is normal to observe a reduction in the early growth of corn sown after oilseed radish (from allelopathy). This effect disappears later and has little influence over yield. It is recommended that it be sown in April to May. Late sowings cut short the vegetative cycle (photoperiod) and therefore produce less biomass.

Oilseed radish may be sown manually by broadcasting the seed and incorporating it superficially with an animal drawn disc harrow (regulated to roll over the ground, cutting as little as possible into the soil), a knife-roller, a rake of branches, etc. It germinates well when the seeds are not covered, as long as soil humidity is maintained by drizzle and/or cloudy days, and also when the soil has sufficient dead cover. In pure sowings it is recommended that 20 kg/ha of seed be used, reducing this quantity when sown in association with other green manure/cover crops. The weight of 1,000 seeds is from 6 to 14 g.

Management to flatten it may be done with a knife-roller, log, machete, etc., 10 to 15 days before the next crop. The optimal moment is when the plants are flowering and the siliques (fruits) begin to fill, about 120 days after sowing. If flattening is delayed and viable seeds remain, these will germinate and may become weeds in the following crop, although they won't develop successfully in hot weather.

Seed production: On small farms it is possible to leave part of the green manure/cover crop for seed production. However, information exists that shows greater yields when lower sowing densities (10 kg/ha) are utilized. Oilseed radish crosses easily with wild radish (Raphanus raphanistrum), therefore fields for seed multiplication and neighboring areas should be free of this weed. The seeds mature gradually between 150 and 180 days after sowing. It is recommended that harvest take place in October when the seeds are totally mature (very dry), cutting the plant with a machete in order to thresh the siliques. Yields range from 300 to 500 kg/ha. A seed-producing area of 700 m^2 is needed to harvest enough seed to sow 1 ha of green manure/ cover crop.

Hairy vetch (Vicia villosa Roth)

Characteristics: It is an annual, herbaceous, creeping, climbing legume of temperate and subtropical climates. It has the capacity to develop in acid soils with aluminum present. It produces about 3 t/ha of dry matter on soils of medium fertility. It doesn't develop well on extremely degraded soils, producing in this case less than 2 t/ha of dry matter. It displays slow initial development, but because of its growing habit (climbing stems and creeping growth) achieves good soil cover and smothers weeds.

Ways to use: Hairy vetch is preferably sown in April, following summer crops (cotton, corn, sesame). Because of its long cycle (7 to 8 months), it is an interesting alternative to reach October-November with good soil cover and few weeds. This facilitates No-Till sowing of cotton, sesame, corn, or other late-sown crops. Another alternative is to associate hairy vetch with out-of-season corn or cassava, since it may be sown from April until July. It is preferable that hairy vetch be mixed with black oats or rye. On small farms, it is recommended that it be sown broadcast, using 50 to 60 kg/ha of seed for pure sowings. The weight of 1,000 seeds is about 25 to 38 g. For good germination, it is necessary that seeds be in contact with the soil; this is achieved by passing a knife-roller, disc harrow (regulated to roll over the ground, cutting as little as possible into the soil), etc.

When one wishes to sow crops over hairy vetch cover, it should be flattened 10 to 15 days before sowing. Management to flatten hairy vetch may be carried out until October, desiccating it with an herbicide, or opening furrows (hoe or ripper), or opening narrow paths (hoe, machete, herbicide, etc.). In these two last cases, the hairy vetch will continue to develop between rows. After October, its cycle will end or it will die from the heat.

Seed production: It is necessary to dedicate a plot exclusively for seed production, because only a small quantity is produced at the sowing densities used for a green manure/cover crop. Furthermore, it has a very long cycle that could delay the sowing of the following crop. In order to produce seed, it is recommended that 25 to 30 kg/ha of seed be used. It is preferable that hairy vetch have a support, which could be a grass such as oats, sorghum, or corn, or an erect legume such as white lupine or pigeon pea. This way, hairy vetch will climb and have more light and air circulation, increasing flowering and yields. Seeds mature in November-December; its cycle is completed at 200 to 240 days. Hairy vetch seedpods have a high incidence of dehiscence (they split open), therefore should be harvested when 60 to 70% of seedpods are dry. A delay in harvest could imply an important loss of seed. Yields may range from 300 to 600 kg/ha in a pure crop, up to 700 to 1,000 kg/ha in a crop with support. A seed-producing area of 2,000 m^2 is needed to harvest enough seed to sow 1 ha of green manure/cover crop. Seeds should be stored under cool conditions.

Rye (Secale cereale L.)

Characteristics: It is an annual, erect grass of medium height. Among grains, it is considered to be the species most resistant to cold and frost. It is drought tolerant. It does not withstand waterlogged soil, but develops well in humid soils with good drainage. It's capable of growing on acid soils, either sandy or clay. In soils of medium and high fertility it displays good growth (around 3.5 to 4 t/ha dry matter). It does not develop well on extremely degraded soils where it produces less than 2 t/ha of dry matter. It responds well to chemical and organic fertilization, and competes well with weeds.

Ways to use: In crop rotations it may be sown in April, preferably to precede a leguminous crop, being an alternative to the use of black oat. It adapts well to broadcast sowing; 80 kg/ha of seed are utilized. The weight of 1,000 seeds is from 15 to 18 g. It may be associated with grasses (oats, ryegrass), legumes (vetch, white lupine, etc.), and crucifers (oilseed radish).

It may be flattened to form cover with a knife-roller 2 to 3 weeks before the following crop, preferably while in the milky grain stage, at which time there is a lower incidence of resprouting (around 130 days after sowing). If sprouting should occur, the recommendation is to desiccate it with an herbicide, or open furrows (hoe or ripper), or make narrow paths (hoe, machete, herbicide, etc.)

Seed production: Seeds may be harvested from the same field sown for a green manure/cover crop. It is important to pay attention to the time of harvest (September-October); rye sheds its grain easily, therefore a delay could cause great losses. Productivity may vary from 800 to 1,500 kg/ha. The crop's cycle is complete in about 140 to 160 days. The seed can be harvested and threshed manually. A seed-producing area of 1,000 m^2 is needed to harvest enough seed to sow 1 ha of green manure/cover crop.

Ryegrass (Lollium multiflorum Lam)

Characteristics: It is an annual, erect grass of medium height. Although it develops in subtropical climates, it is demanding of cold, therefore its greatest potential for utilization is in the colder regions (Departments of Itapúa and Alto Paraná). It grows well on soils of moderate to high fertility, producing 3.5 to 5 t/ha of dry matter. It doesn't grow well on very degraded soils, where it produces less than 3 t/ha of dry matter. It responds well to the addition of nitrogen and phosphate mineral fertilizers, and also to organic fertilizers. It is demanding in terms of humidity, but does not tolerate being waterlogged. Association with other species, such as black oats, can favor its establishment by creating a favorable microclimate, principally in warmer regions. It demonstrates a high capacity for natural resowing in cold regions, which is an advantage in systems where there are no winter crops, or in soils with steep slope that are left fallow. Seeds that fall in the spring germinate the following autumn. In this case, ryegrass can become a weed in winter crops. The harvest and threshing of ryegrass are done manually.

Ryegrass provides good soil cover, and has a great quantity of roots that contribute to soil aggregation. It reduces the population of nematodes. It is a forage of high quality that can be grazed directly by animals, or in the form of hay. Its initial growth is slower than that of black oats, and it is used as forage in winter and spring. Ryegrass favors weed suppression, reducing the need for control in the following crop (fewer hoeings and/or less use of herbicides).

Ways to use: It has a positive effect when it precedes the cultivation of soybeans and other legumes. It is recommended that it be sown in March to April, the early sowings being suitable for the colder regions. Also, early sowings are more appropriate for the production of biomass for cover or forage, and late sowings for seed production.

Ryegrass may be sown broadcast or in rows, and the seed should remain on the soil's surface. In the case of broadcast sowing, contact between the seed and soil can be favored by passing a log or knife-roller, taking care that the seeds not be buried. When sowing depth exceeds 1 cm, the seed generally does not germinate. In pure sowings a density of 25-30 kg/ha is recommended, whereas in association with other gramineous (oats, rye) and/or leguminous (vetch, etc.) species, it is recommended that approximately 17 kg/ha be utilized. The weight of 1,000 seeds is from 2 to 3 g.

The optimal time to flatten ryegrass for cover is at 130 to 170 days after sowing, when it is in full flower; however, it should be flattened according to the sowing date of the following crop (2 to 3 weeks before sowing). The knife-roller alone is not enough to kill ryegrass, and it is necessary to complement it with an herbicide, or to open furrows (hoe or ripper), or to open narrow paths (hoe, machete, herbicides, etc.).

Seed production: For seed production it is advisable to sow late, in plots especially intended for that purpose, at similar densities as when sown for a green manure/cover crop. Seeds may also be harvested from fields of GMCC. Inclusively, it may be grazed and seed harvested later. Ryegrass is harvested in November, 1 to 2 weeks after the milky grain stage in order to avoid loss from seed fall. Good seed quality may be achieved by drying immediately after harvest, which can be done manually. On small farms, seeds may be dried by spreading them in the sun and stirring periodically. Yield per hectare can vary around 600 to 800 kg/ha. The complete cycle is about 210 days. A seed-producing area of 500 m^2 is needed to harvest enough seed to sow 1 ha of green manure/cover crop.

Sunflower (Helianthus annuus L.)

Characteristics: It belongs to the composite family. It is an annual, herbaceous, erect plant of medium to tall stature (reaches a height of 1 to over 3 m, depending on the variety). It has a deep, abundant root system. It is drought tolerant (can grow with 250 to 400 mm of rain), and has the capacity to grow almost all year in Paraguay's climatic conditions. It stands out for its rapid initial growth and good shading of the soil. This, together with its pronounced allelopathic effect, makes it one of the most efficient plants for weed suppression (when sown at high densities), inclusively in short periods of time, from 50 to 60 days. Its dead cover decomposes rapidly and, therefore, the following crop should be sown immediately after it is flattened to avoid weed problems. It is one of the winter green manure/ cover crop species that produces more biomass in soils of medium and high fertility (4.5 to 7 t/ha of dry matter), principally the varieties Guayakán and Peredovic. It does not develop well on extremely degraded soils (less than 2 t/ha). It recycles a great quantity of nutrients through its vegetative mass, but its residual effect on the following crops is not outstanding in relation to other green manure/cover crops.

Due to its susceptibility to several diseases (Alternaria, Rust, Macrophomina, Sclerotinia, and Phomopsis), sunflower should be inserted in an appropriate crop rotation. Furthermore, it doesn't grow well when sown in the same field year after year (self-incompatibility). The recommendation is to wait a minimum of three years before sowing sunflower in the same place.

Ways to use: It may be included between two crops when a short period of time is available ; for example, after out-of-season corn that is harvested in July-August and preceding the cotton crop. Hard frost can damage plants when they are at the height of their growing phase. In addition, sunflower may be utilized in February-April sowings after summer crops such as corn, cotton, peanuts, etc.

It is recommended that varieties such as Guayakán, Peredovik, Estanzuela, etc. be used. Sunflower may be sown with a manual jab planter, pointed stick, etc., at a density of 20 to 25 kg/ha of seed, spaced at 50 cm between rows with 30 cm between holes and 5 to 6 seeds per hole. Sowing depth should be from 3 to 5 cm. The weight of 1,000 seeds of the variety Peredovic varies from 70 to 120 g. It is preferable that sunflower be associated with other GMCCs, such as black oats and hairy vetch, to improve cover; in this situation greater spacing may be used.

Management to flatten sunflower may be done immediately before the following crop, with a knife-roller or machete, preferably while in full flower (90 to 120 days after sowing). At this time it reaches maximum biomass production. When used as a short-term GMCC, sunflower may be flattened at 50 to 70 days.

Seed production: Fields of green manure/cover crop may be utilized for seed production. Ideally, sunflower is harvested when the achenes (grains) are mature and dry. However, due to the high risk of loss to birds, it's possible to harvest ahead of time if the seeds are immediately dried in the sun to avoid problems with fungi and other diseases that can reduce germination and vigor. It may be harvested manually in September-October. The crop's cycle fluctuates around 150 days, depending on the variety and time of sowing. Seed yields may reach approximately 1,000 kg/ha or more. A seed-producing area of 250 m^2 is needed to harvest enough seed to sow 1 ha of green manure/cover crop.

7

Economics of Conservation Agriculture

Conservation agriculture (CA) aims to make better use of agricultural resources through the integrated management of available soil, water and biological resources, combined with limited external inputs. It contributes to environmental conservation and to sustainable agricultural production by maintaining a permanent or semi-permanent organic soil cover. Zero or minimum tillage, direct seeding and a varied crop rotation are important elements of CA.

Adoption of CA at the farm level is associated with lower labour and farm-power inputs, more stable yields and improved soil nutrient exchange capacity. Crop production profitability under CA tends to increase over time relative to conventional agriculture. Other benefits attributed to CA at the watershed level relate to more regular surface hydrology and reduced sediment loads in surface water. At the global level, CA sequesters carbon, thereby decreasing CO2 in the atmosphere and helping to dampen climate change. It also conserves soil and terrestrial biodiversity.

Conservation agriculture is practised on about 57 million ha, or on about 3 percent of the 1 500 million ha of arable land worldwide. Most of the land under CA is in North and South America. It is rapidly expanding on small and large farms in South America, where practising farmers are highly organized in local, regional and national farmers' organizations. In Europe, the European Conservation Agricultural Federation, a regional lobby group, unites national CA associations in the United Kingdom, France, Germany, Italy, Portugal and Spain.

Despite these apparent advantages, and despite the few notable exceptions in the developing world, CA has spread relatively slowly, especially in farming systems in temperate climates. The transformation from conventional agriculture to CA seems to require considerable farmer management skills and involves investment in new equipment. However, it may also require minimum levels of social capital to foster its expansion.

Defining Conservation Agriculture

CA has emerged as an alternative to conventional agriculture as a result of losses in soil productivity due to soil degradation (e.g. erosion and compaction). CA aims to reduce soil degradation through several practices that minimize the alteration of soil composition and structure and any effects upon natural biodiversity. In general, CA includes any practice that reduces, changes or eliminates soil tillage and avoids the burning of residue in order to maintain adequate surface cover throughout the year. In contrast, conventional forms of agriculture regularly use ploughs to enable a deep tilling of the soil.

The line between conventional and CA often blurs as conventional agriculture utilizes many practices typical of CA, such as minimum or no-tillage. Hence, the differentiating feature of CA and conventional agriculture is the mind-set of the farmer. The conventional farmer believes that tilling the soil will provide benefits to the farm and would increase tillage if economically possible. On the other hand, the conservation farmer questions the necessity of tillage in the first place and feels uncomfortable when tillage occurs.

CA maintains a permanent or semi-permanent organic soil cover consisting of a growing crop or a dead mulch. The function of the organic cover is to physically protect the soil from sun, rain and wind and to feed soil biota. Eventually, the soil micro-organisms and soil fauna will take over the tillage function and soil nutrient balancing, thereby maintaining the soil's capacity for self-recuperation. Residue-based zero tillage with direct seeding is perhaps the best example of CA, since it avoids the disturbance caused by mechanical tillage. A varied crop rotation is also important to avoid disease and pest problems. The last two decades have seen the perfecting of the technologies associated with minimum or no-tillage agriculture and their adaptation for nearly all farm sizes, soil and crop types and climate zones.

Some examples of CA techniques include:

— Direct sowing/direct drilling/no-tillage: The soil remains undisturbed from harvest to planting except for nutrient injection. Planting or drilling takes place in a narrow seedbed or slot created by coulters, row cleaners, disk openers, in-row chisels or roto-tillers. Weed control is primarily by herbicides with little environmental impact. Cultivation is a possibility for emergency weed control. This strategy is the best option for annual crops.

— Ridge-till: The soil remains undisturbed from harvest to planting except for nutrient injection. Planting takes place in a seedbed prepared on ridges with sweeps, disk openers, coulters, or row cleaners. Residue is left on the surface between ridges. Weed control is by herbicides and/or cultivation. Ridges are rebuilt during cultivation.

— Mulch till/reduced tillage/minimum tillage: The soil is disturbed prior to planting. Tillage tools such as chisels, field cultivators, disks, sweeps or blades are used. Weed control is by herbicides and/or cultivation. In non-inversion tillage, soil is disturbed (but not inverted) immediately after harvest to partially incorporate crop residues and promote weed seed germination to provide soil cover during the intercrop period. These weeds are later chemically destroyed (using herbicides) and incorporated at sowing, in one pass, with non-inversion drills.

— Cover crops: Sowing of appropriate species, or growing spontaneous vegetation, in between rows of trees, or in the period of time in between successive annual crops, as a measure to prevent soil erosion and to control weeds. Cover-crop management generally utilizes herbicides with a minimum environmental impact.

The definition of CA used in this study is broader than that used by FAO (no-tillage with direct seeding and maintenance of soil cover/crop residues with no incorporation, along with crop rotations). The wider interpretation of the concept encompasses a larger number of data and informational sources, as many studies employ differing definitions of CA and the broad definition presented here captures most of this variation.

Economic Rationale for Promoting Conservation Agriculture

The distinction between local, national and global impacts is important as it is possible to rationalize national or global programmes supporting the adoption of CA according to how significant the net benefits are at this level. The benefits at the national level are especially important and they strongly argue for policy support at this level. Uri *et al.* estimated that the realized erosion benefits (avoided losses from sheet, rill and wind erosion) for the United States from the existing areas under conservation tillage ranged from US$90.3 million to US$288.8 million in 1996.

From the farmer's perspective, the benefits of CA can be either on-site (private) or off-site (reduced sediment pollution, carbon sequestration, etc.).

Few empirical studies consider the economic benefits of adopting CA in the tropical agro-ecological zone, so most accumulated evidence is for developed regions such as North America. For example, Stonehouse simulated full-width no-plough and no-till use in southern Ontario, Canada, and found that both provided modestly higher on-farm benefits than did conventional tillage. The advantage of no-plough and no-till was even greater with off-site benefits included. The off-site benefits considered were downstream fishing benefits and reduced dredging costs. These accounted for 43 percent and 10 percent, respectively, of the net social benefits from conservation tillage. Thus, despite marginally higher profits under CA, the inability to capture off-site benefits means that fewer farmers adopt CA than might otherwise be the case.

Other studies find a trade-off between economic returns and environmental integrity with the adoption of increasingly intensive conservation agricultural practices. Kelly *et al.* find that strict no-till produces higher returns than conventional tillage and reduces an environmental hazard index from 78.9 to 64.7. The index takes into consideration soil erosion risk, phosphorous and nitrogen losses, and potential pesticide contamination. By further incorporating cover crops and replacing fertilizers with manure, the CA option becomes less profitable than conventional tillage. However, the environmental hazard index declines to 50 or lower, making the economic-environmental trade-off clear from a social perspective.

The global concern about soil degradation helps support an argument for intervention at the international level. This argument stems not just from a concern about what is occurring within individual nations but also from the possible presence of regional or global costs imposed by soil degradation. In other words, there may be global benefits from adopting CA and other soil-enhancing technologies.

There are potential global benefits associated with the adoption of CA. For example, there is a link between carbon sequestration in soil and global warming as the long-term capture of carbon in organic matter reduces the atmospheric load of carbon. However, the benefits associated with carbon sequestration in soil may be elusive if soil degradation results in a transfer of carbon from one location to another with no net release to the atmosphere. For CA, Uri (1999a) argues that the "benefits to be gained from carbon sequestration will depend on the soil remaining undisturbed".

In the absence of sustainable soil management practices, soil degradation can lead to crop and livestock losses, with regional or global consequences (refugees, famine, etc.). Where the rest of the world provides assistance, these resources are wasted if the earlier adoption of CA or other practices could have avoided the situation. In addition, lands under CA support terrestrial wildlife and soil microfauna that are important components in global biodiversity, as demonstrated by the discovery of penicillin and streptomycin. Thus, good soil conservation and management can have benefits that the individual farmer does not anticipate, but which do have real implications for the global environment.

A Conceptual Framework for Studying the Adoption of Conservation Agriculture

Farmers who switch to some new technique from conventional practice may do so for a variety of reasons. They may detect a more efficient and profitable way to produce, or they may perceive a problem and in seeking solutions arrive at a new practice, such as CA. The problems stimulating the possible change to CA are typically soil degradation, soil erosion or declining crop yields due to deteriorating soil fertility. These views are associated with the traditional model of innovation and the adoption of new technologies in many industries, including agriculture.

Some farmers have adopted CA because they found that immediate yield benefits or profits were attractive. In this situation, a clear financial incentive has induced the change in behaviour. However, it may be inappropriate to rely on the classical model as a basis for promoting the adoption of agricultural conservation technologies (e.g. no-till). This is because the adoption and diffusion model is based on "voluntarism on the part of the farmer's decision making and the economic gain attached to the new behaviour".

Moreover, some authors argue for the presence of a continuous complex innovation process governing agricultural technologies such as CA, using the example of zero tillage. These innovation systems are non-linear and involve complex interactions and feedbacks among agents (e.g. farmers, extension agents, and private enterprises). These authors argue that continuous complex innovation systems are characterized by the presence of agents that have limited information but are always in search of new technological opportunities. In addition to individual agents' actions, initial circumstances and the working of feedback loops have a great bearing on the innovation process, making it unpredictable. The resulting technological innovation stems from a particular mix of initial conditions, random events and long-term trends. As an example, the response of pests to new control techniques is unpredictable, yet has a significant influence on the evolution of future technology development and adoption.

Regardless of the motivating factor or the model of adoption assumed, farmers consider only those aspects of their operation that are relevant from a private perspective. This process typically involves only on-farm considerations. However, it could extend to impacts on neighbours and future generations if social relations and stewardship considerations receive high personal priority. Despite the more limited view, many factors influence this private perspective and help to mould decisions about new technologies or a change in farm practices..

Information about new technologies and financial conditions is a precursor to changes in farm practices and acquiring it does not usually involve large financial outlays. Government credit and extension policies play an important role here. In contrast to the more direct working of agriculture sector policies and financial incentives, some social and institutional factors have a more indirect influence. Nonetheless, all these factors affect the net returns, risks and other pecuniary elements that drive the decision-making process.

Central to this model of the decision-making process are farmers' perceptions. Changing policy and financial incentives or declining natural resource quality signal to the farmer that the current pattern of use of household resources may no longer be desirable. There is controversy over the extent to which farmers perceive progressive deterioration in their natural resource base. However, there is now sufficient evidence that smallholders are often aware of soil degradation, although other factors affecting production may mask this at times.

CA is just one of many options available to farmers responding to perceived changes in their production environment. For example, all or a few of the household's members may migrate or accept off-farm employment, or remain behind and modify farming practices. Critically, the impact on soil productivity can be either positive or negative, depending upon numerous factors. If households choose migration, they may reduce the intensity with which they farm existing plots, or abandon their old lands altogether and bring new land in frontier areas under cultivation. The latter can have serious implications if farmers transfer unsustainable soil management practices to new areas. There are also many technical alternatives available to producers if they choose to change existing management rather than migrate, and these include CA. The choices of individual farmers are cumulative and can have eventual impacts well beyond the individual farm.

Financial analyses of conservation agriculture versus conventional practices

It might be assumed that CA is more profitable in steep-sloping, high rainfall tropical regions (e.g. Latin America) than in flatter temperate areas (e.g. Canada, the United States), since the former would be subject to a higher risk of erosion under conventional tillage. But such a generalisation would hide a number of the complexities that make the analysis of financial returns from CA difficult. For example, in 7 of the 12 recent cost studies reviewed for this study, reduced or no-tillage showed higher net returns than conventional tillage, and most of these studies involved temperate regions.

The temperate agro-ecological zone in developed countries

One of the first comprehensive financial analyses of CA on large farms in developed countries compared the on-farm costs of conventional tillage with conservation tillage in the United States. More recent reviews have tended to reinforce its conclusion that CA has a small cost advantage over conventional tillage but that site-specific conditions could alter this result in various ways. The following input cost aspects form the basis for these general conclusions.

Machinery and fuel costs

This is the most important cost item for larger producers and so the impact of CA on these expenditure items is critical. Most analyses suggest that CA reduces machinery costs. Zero or minimum tillage means that farmers can use a smaller tractor and make fewer passes over the field. This also results in lower fuel and repair costs. However, this simple view masks some complexities in making a fair comparison. For example, farmers may see CA as a complement to rather than as a full substitute for their existing practices. If they only partially switch to CA (e.g. on some fields or in some years), then their machinery costs may rise as they must now provide for two cultivation systems, or they may simply use their existing machinery inefficiently on their CA fields.

To capture such complexity, economists distinguish between short-run and long-run costs, where the former assumes no adjustment to existing capital equipment and the latter assumes such an adjustment. A comparative study of CA and conventional tillage in Wisconsin found that short-run average costs under CA exceeded long-run average costs by about 7 percent. The short-run average costs per hectare for CA were greater than for conventional tillage. However, after adjustments to capital, CA costs fell below those of conventional tillage in the long run.

Similarly, the expectation is for fuel costs to be lower under CA, and this is generally the finding in most studies. Falling fuel prices should encourage greater adoption of CA. One study shows that the price of crude oil has a statistically significant but relatively minor effect on the intensity of CA (but not adoption by new farmers). It finds that a 10 percent increase in the United States in the price of crude oil is associated with an expansion in planted hectares under CA of 0.4 percent, with the expansion being concentrated primarily on existing CA farms.

Pesticide costs

Offsetting lower machinery costs are higher herbicide applications under CA, especially during the early adoption period and with no-till. Indeed, herbicides substitute for the use of machinery to keep weeds under control. Site-specific factors are important as perennial weeds can present problems for CA. Nonetheless, herbicide application rates and the ability to fully control weeds under CA in all situations remains a controversial and continuing area of CA research. Recent assessments have tended to argue that herbicide applications decline over time and may eventually fall to a level equal to that of conventional tillage. Insect control is less an issue in conventional and CA comparisons. As most pesticides are petroleum based, crude oil prices are liable to affect their cost to farmers. If so, then a higher crude price would mean higher herbicide costs, partially offsetting CA's relative cost advantage stemming from lower machinery fuel requirements.

Labour costs

Much attention has focused on the apparent reduction in labour requirements under CA. This reduction follows from the decreased demand for labour for land preparation at the beginning of the growing season. Some estimates put this reduction at 50-60 percent during this time period. On large mechanized farms in the developed world the true impact of this saving is small as labour costs account for under 10 percent of total per acre costs. However, on some farms in the developed world, the trend towards increased off-farm work has made even the relatively small labour savings under CA attractive. Indeed, some case studies have cited the time savings provided by CA as the primary motivation for the adoption of conservation tillage.

Fertilizer and other input costs

Most comparative analyses of the costs of conventional tillage versus conservation tillage assume that other production inputs remain unchanged following a switch to CA. A debate continues concerning fertilizer use under CA as there is evidence, that CA adoption affects nitrogen use by crops and leaching. Uri finds some increase in fertilizer use by maize farmers adopting conservation tillage in the United States. Additionally, if the application of fertilizers under CA requires greater management skill, then application costs could rise even if application rates do not. A more general finding is that CA requires greater management skills and it may be costly for farmers to acquire these. CA may also affect seed purchases as farmers may be able to avoid some pest problems by investing in more resistant seed varieties. However, this increases costs.

Recent estimates tend to show a wide range for CA, recognizing the variation in site-specific conditions (e.g. drainage, rainfall). Putting the cost savings attributable to CA in the context of these total costs, any cost advantage amounts to about 5-10 percent in 1979 and probably about the same in the 1990s.

Also missing from many cost comparisons of conventional and conservation tillage is an analysis of risk factors. One aspect of risk is a recognition that yields might vary under the different cultivation systems. Much debate has centred on whether switching to CA leads to higher or lower yields. As the results for temperate climates are often contradictory, and any differences are usually not statistically significant, most analysts simply assume no change in yield. Similarly, the impact of adopting CA on yield variability and risk is controversial. Some studies argue that CA increases yield variability in many situations, thereby worsening risk. By contrast, Australian research shows a reduced variability in crop yields with CA, while work in Canada indicates that the net returns were higher under CA than conventional practices in bad years, but lower when averaged over time. Firm conclusions on whether risk is increased or reduced under CA remain elusive.

More certain are the impacts of CA on cropping intensity. With reduced field preparation time, the cropping cycle is shorter, allowing more crops in a given period and even double cropping where it was not possible previously. Where this benefit is available from CA, more efficient utilization of the fixed land resource results in higher annual net returns per hectare. Moreover, farmers may adjust their cropping strategy when switching to CA. Hence, yield trials comparing the same crop under either cultivation system may not represent reality. In fact, fully adopting CA involves switching to a suitable crop rotation that will probably differ from the conventional cropping strategy used previously. For this reason, some writers have called for a broader whole farm approach to comparative assessments in temperate agriculture.

Overall, a comparison between conventional and conservation practices in temperate agro-ecological zones hinges on two offsetting effects. One involves CA's labour and possibly machinery cost savings, while the other involves higher herbicide costs, at least initially, under CA. Depending upon the magnitude of each of these effects, CA may appear either more or less costly. For example, in Saskatchewan, Canada, researchers found that the higher herbicide costs characterizing CA overwhelmed any cost savings associated with labour, fuel, machine repair and overheads. Similarly, Stonehouse and Bohl used a linear programming model to argue that conservation tillage in a cash-crop farm system is not profitable. However, most developed-country studies reviewed find that CA demonstrates at least minor cost savings over conventional practices. However, these savings have not been sufficient to induce adoption by large numbers of farmers on large mechanized farms. These farmers may resist new practices unless there is a promise of much higher financial returns.

Tropical/Remperate Agro-ecological Zone

One of the success stories for CA has been in Latin America. Large-scale mechanized farming is common in many parts of Latin America and farmers have adopted CA on large portions of this cultivated area. While most of the comparative cost analysis presented above for temperate northern regions would apply here, the advantage of CA in Latin America has been more pronounced. In part, this greater advantage reflects physical and climate factors, but also the differences in the nature of the technology adopted.

In Paraguay, yields under conventional tillage declined 5-15 percent over a period of ten years, while yields from zero tillage increased 5-20 percent (Sorrenson*et al.*, 1997 and 1998). Savings in fertilizer and herbicide inputs dropped by an average of 30-50 percent over the same period. In Brazil, over a 17-year period, maize and soybean yields increased by 86 and 56 percent, respectively, while fertilizer inputs for these crops fell by 30 and 50 percent, respectively. In addition, soil erosion in Brazil fell from 3.4-8.0 t/ha under conventional tillage to 0.4 t/ha under no-till, and water loss fell from approximately 990 to 170 t/ha.

As a result, the financial benefits for farmers in Latin America who have adopted CA have been striking. However, these take time to fully materialize. Sorrenson compared the financial profitability of CA on 18 medium and large-sized farms with conventional practice in two regions of Paraguay over 10 years. He found that by the tenth year net farm income had risen on the CA farms from under US$10 000 to over US$30 000, while on conventional farms net farm income fell and even turned negative.

Medium and large-scale farmers have experienced:

- less soil erosion, improvements in soil structure and an increase in organic matter content, crop yields and cropping intensities;
- reduced time between harvesting and sowing crops, allowing more crops to be grown over a 12-month period;
- decreased tractor hours, farm labour, machinery costs, fertilizer, insecticide, fungicide and herbicide, and cost savings from reduced contour terracing and replanting of crops following heavy rains;
- lower risks on a whole-farm basis because of higher and more stable yields and diversification into other cash crops.

In Latin America and in other developing regions, CA is a technology with potential appeal for smallholders. However, adopting CA on a small, possibly non-mechanized, farm involves some different considerations when compared to a large mechanized farm. For example, as smallholders use few purchased inputs, discussions on large increases in herbicide costs may not be relevant. Even if smallholders accept the need for herbicides, they may be unable to finance their purchase. In addition, few smallholders use significant amounts of fertilizer so that a debate over the impact of CA on fertilizer use is largely irrelevant. Ultimately, the availability of credit to assist with CA's increased need for purchased inputs plays an important role. If smallholders hire land preparation equipment, then a switch to CA should be relatively simple as there are no machinery investment implications. Short-run costs would be close to long-run costs when switching to CA.

The majority of smallholders worldwide do land preparation and weeding manually, and adopting CA has its greatest impact on the labour used in these activities. In a comparative analysis of traditional bush fallow systems with no-till and alley cropping in Nigeria, labour savings under the no-till technology were substantial (Ehui *et al.*, 1990). Whereas alley cropping required from 126 to 151 person-days/ha/year and the bush fallow system needed from 67 to 102 days, the no-till technology required 58 days (with an allowance for land clearing in each case). These labour inputs amounted to more than 50 percent of total production costs for each technology. However, higher herbicide and equipment costs penalized the no-till technology and it was only preferred under conditions of higher population pressure, which penalizes alternative fallow systems. In studies of smallholders in Latin America, net farm income and returns to labour were much higher under CA than conventional practice.

In judging the attractiveness of CA in smallholder systems in Africa, Latin America and elsewhere, labour savings are a key factor. A further point related to labour is that as the labour savings come at both the land preparation and weeding stages (assuming herbicide use), there are liable to be implications for the gender division of labour. In most smallholder systems in Africa, male household members are responsible for land preparation (with a contribution to sowing), while female household members are responsible for weeding. Herbicide use may require some adjustment in these responsibilities as male household members usually handle pesticides. Male household members may resist the additional labour demand during the weeding period, so creating a barrier to the adoption of CA.

Furthermore, certain conditions can enhance the relative financial attractiveness of CA. For example, rising land pressure tends to increase the attractiveness of CA relative to bush fallowing. An additional consideration is land quality. Studies of the net returns from mulching, an important component in smallholder CA, suggest that the benefits of this practice increase with the quality of cropland. Successful instances of CA adoption in Latin America have demonstrated the importance of credit as an important enabling factor. This is because of the need to finance specialized planting equipment and herbicides.

Financial Analyses of Conservation Agriculture versus Other Conservation Technologies

Most financial analyses of CA concentrate on a comparison with conventional practice, whether this is conventional tillage or bush fallow. However, farmers can often select from a number of alternative conservation practices, in which case CA is just one option of perhaps several. This is especially true for smallholder systems as an absence of prior machinery investments and the small-scale adaptability of many soil and water conservation techniques makes adoption relatively easy in physical and financial terms.

To consider CA's attractiveness in relation to alternative conservation practices to a smallholder, a database of over 130 different analyses of individual soil and water conservation technologies was compiled. The analyses concentrated on Africa and Latin America with all technologies coded according to whether they constituted a CA-related technology (Group 1) or not (Group 2), as specified by the World Overview of Conservation Approaches and Technologies (WOCAT) technology classification system.

Group 1 includes measures aimed primarily at enhancing soil cover and organic matter, while Group 2 technologies are generally linear, cross-slope approaches intended to reduce erosion from wind or runoff. Information about farm-level financial returns was entered in the database for each technology. The results for each of the two technology groups were sorted based on whether technology adoption provided a positive or negative net present value (NPV).

The classification of technologies is not precisely consistent with the definition of CA presented earlier. CA and, more broadly, agronomic improvements tend to show higher net returns at the farm level than do other techniques (e.g. vegetative, structural and other improvements). Arguably, this relative attractiveness of CA is more pronounced than was the case from the comparison of only CA and conventional tillage. Thus, when faced with numerous alternatives to conventional practice, CA and related approaches may offer the best possible returns in many situations. Site-specific factors would determine which individual technology offered the best returns for individual farmers.

In summing up the financial evidence in support of CA, a few words of caution are in order. While it is true that CA often conforms to what Pampel and van Es term an 'environmentally profitable practice' (i.e. good for environment and profitable), this is not always so. Particular location constraints might result in reduced yields, or institutional factors may favour alternative practices.

Thus, it is necessary to consider site-specific conditions in determining the financial attractiveness of CA. Even where the financial incentives may appear attractive, a consideration of non-financial factors is required to understand the actual and potential adoption of CA.

Other Factors Influencing the Adoption of Conservation Agriculture

A number of studies have sought to identify barriers to adoption beyond the obvious divergence between on-farm costs and wider social benefits under CA. For example:

— Large investment costs may discourage adoption.

— The perceived risk of adopting CA may serve as a barrier.

— Long gestation periods for the benefits of CA to materialize may serve as a barrier to farmers with short-term planning horizons.

— Barriers may be particular to culture and recent history.

In part, the need to consider factors other than net returns reflects farmers' competing objectives in farm management, i.e. profitability versus low investment or minimum subsistence food requirements. Competing technologies may meet individual objectives to varying degrees. In terms of maximizing net financial returns, CA can provide better net returns than either conventional practice or other conservation technologies, subject to local site conditions.

Farm-level factors vary from farm operation to farm operation and higher level factors are also at work, such as the transmission of information (via policy-related activities and social processes). Furthermore, the variables discussed below, and their broader categories, do not act independently, but rather interact to influence adoption.

Farmer Characteristics

Since Ryan and Gross first showed that the adoption of agricultural innovations is typically uneven from farmer to farmer, researchers have directed attention to certain characteristics and attributes of farmers in an effort to explain this unevenness. In the case of soil conservation technology adoption, Gould*et al*. emphasize awareness on the part of farm operators to soil erosion or other soil problems as an obvious prerequisite to adoption. Indeed, farmer awareness or perception of soil problems is frequently found to positively correlate with CA adoption. Similarly, the central place of information and knowledge in

CA adoption, in terms of being aware of soil problems and potential solutions, should lead the level of education of a farm operator to correlate positively with adoption. Education, be it specific or general, generally correlates positively with the adoption of CA practices, notwithstanding some findings of insignificance or even negative correlation.

Age and/or experience are difficult factors to link to CA adoption, given that studies have shown both a positive and negative correlation. Based on a study of conservation tillage adoption in Wisconsin, Gould *et al.* showed that older and more experienced farmers were more likely than their younger colleagues to recognize soil problems. However, they were less likely than their younger colleagues to address the problems once recognized. In contrast, several studies have found that income correlates positively with the adoption of soil-erosion control practices.

Farm characteristics

Studies of the adoption of conservation tillage and other CA-type practices have often given significant attention to farm size (or sometimes planted area). Many studies have found that farm size correlates positively with adoption. However, other studies have shown no significant relationship, or even a negative correlation. Hence, the overall impact of farm size on adoption is inconclusive.

Some studies have found that the presence of soil erosion and other soil problems on the farm correlates positively with conservation tillage adoption. However, farmer awareness of and concern for soil problems is probably the more critical factor affecting adoption. Another important farm characteristic is underlying land productivity. In the case of no-till and mulch tillage, Uri shows that in the United States adoption is more likely on farms with low rather than high levels of soil productivity. In addition, a good fit between CA and the farm's production goals encourages adoption.

A more complex factor liable to affect adoption is land tenure. In simple terms, privatizing land should lead to better incentives for the adoption of conservation technologies. However, studies of the privatization of land or titling have not shown that this is necessary to motivate sustainable practices and, in some instances, it has had the opposite effect. As a result, it appears that producers may accept titling because it guarantees land rights, but this does not necessarily bring about changes in their land management. In contrast, there are numerous studies indicating that traditional institutions governing access to land resources in developing regions are flexible in responding to internal and external pressures.

Information

Without knowledge of the practices associated with CA via some information or communication channel, adoption is improbable. Indeed, studies of innovation adoption and diffusion have long recognized information as a key variable, and its availability is typically found to correlate with adoption. Information becomes especially important as the degree of complexity of the conservation technology increases.Information sources that positively influence the adoption of CA-type practices can include: other farmers; media; meetings; and extension officers. However, with respect to this latter source, Agbamu shows that contact alone will not promote adoption if information dissemination is ineffective, inaccurate or inappropriate. Studies have not always shown that the ease of obtaining information correlates with adoption.

Biophysical and technical factors

In technical terms, the characteristics and availability of CA technologies are crucial factors in adoption. However, de Harrera and Sain note that availability does not imply individual ownership of the necessary machinery as lease/hire arrangements proliferate. Furthermore, potential adopters must believe that the technology will work. Technical factors interact with biophysical factors, e.g. soil type, rainfall or topography can encourage/facilitate or discourage/limit CA adoption. While some studies have shown that farm operations located within regions of steep slopes and erodible soils have a greater tendency to use CA practices, other studies have found these variables to be insignificant.

Social Factors

CA adoption is seldom strictly a function of individual profit maximization alone, but also can reflect non-individual or societal interests. More specifically, Lynne argues that farmer decision making usually reflects a compromise between private economic utility and collective utility. Producers often identify this latter interest as 'the right thing to do', at least in those places where stewardship is part of the cultural norm.

The argument runs that for many producers the pride associated with stewardship makes up for limits in financial rewards. Examples of such stewardship motives governing land management arrangements include the Landcare movement in Australia.

In contrast, Van Kooten *et al.* modelled the trade-offs between stewardship and net returns on wheat-fallow farms in Saskatchewan, Canada. Their study found that farmers make improvements in agronomic practices to benefit soil quality only under extreme degrees of concern (e.g. stewardship). This result holds despite such practices representing no more than a 5 percent sacrifice in net returns.

In addition to stewardship motives, collective action may be necessary to implement CA on a regional basis. Cooperative arrangements govern numerous activities within village agricultural systems. Although the discussion usually focuses on common property resources, even private land use may overlay with cooperative arrangements governing various aspects of farm management. For example, contour ploughing, stone lines and other structural works require cooperation amongst several or many farmers in order to be effective conservation strategies. Many dimensions of CA fit the cooperative model, including the formation and operation of farmers' groups, dissemination of information, pest control and the purchase of agrochemical inputs.

If CA requires collective action or high levels of social organization to help it gather momentum, then widespread adoption may be related to a society's social capital. The role of social capital in fostering or retarding the collective action needed in promoting new conservation technology is of growing interest. In the broadest sense, social capital refers to the interconnectedness among individuals in society and considers relationships as a type of asset. Several studies have examined the influence of social capital on technology adoption in either developed or developing countries. For example, kinship, or more exactly 'connectedness to others', can influence the adoption of conservation technology. Some studies have shown that the expectation of farmland inheritance can have a bearing on conservation behaviour amongst farmers, although other studies testing for this have not shown a positive correlation. Similarly, higher levels of social capital help explain the adoption of fertilizer and soil conservation practices in Peru, while one study has related the success of peasant committees in Paraguayan villages to the level of social capital in these communities. Such institutions at the local level have been an important catalyst in the adoption and diffusion of CA.

In conclusion, the inconsistent and sometimes contradictory results obtained from studies of the adoption of CA-type practices tend to suggest that the decision-making process is highly variable, and that outcomes may be specific to particular people, places and situations. This makes the task of developing a policy framework to promote CA adoption particularly challenging.

The Influence of Policy on the Adoption of Conservation Agriculture

Agriculture has been subject to considerable state interest and intervention over the past half-century, perhaps more than any other economic sector. While it is possible to overestimate the influence of policies in farmer decision making, there is increasing recognition that the provision of public support in the form of guaranteed output prices, input subsidies, deficiency payments, cheap credit, or disaster relief has encouraged and facilitated massive investment by farmers in production capacity expansion. Some authors have characterized the resulting dominant form of agriculture, at least in the developed world, as industrial. This is because of its continuing trend towards larger and fewer units of production, regional and enterprise specialization, more intensive soil tillage, increased reliance on agrochemicals, and in many locations, surplus output. Given its associated effects upon the quality of soil, water and wildlife habitat, various authors have implicated agriculture policy as a contributing cause of environmental degradation.

It is in this context that many governments have introduced a variety of programmes to encourage the adoption of CA-type practices. With extension services, subsidies and taxes, these initiatives have achieved some important results. For example, the success in promoting CA practices in certain developing regions, particularly Latin

America, is noteworthy, and policy has played an important role.

Many programmes promoting CA throughout the world have been relatively ineffective because of contradictory signals and incentives from existing subsidy programmes. For example, policies designed to promote sustainable agriculture can be undermined by other, typically richer, policy measures in support of highly erosive row crops such as groundnuts and tobacco, or by weak or slow-to-respond research and extension efforts.

Some studies have shown government-financed extension to have a positive impact on adoption, although Agbamu cautions that not all forms of extension will achieve such an end. In the case of state financial assistance, Napier and Camboni identify a positive, albeit weak, correlation between participation in such programmes and conservation tillage adoption.

More specifically, based on a model cash crop farm in southwest Ontario, Stonehouse and Bohl show that a one-time subsidy covering 20 percent of the outlay costs would induce a farmer to convert from conventional tillage to no-till. However, the study suggests that conversion to permanent cover crops such as alfalfa would require excessively high subsidies.

Finally, with respect to the use of taxes, Aw-Hassan and Stoecker determined that if the off-site damages from conventional practices were taxed as high as US$2.25 per tonne of soil loss, the area of high-yielding/high-erosion land under conservation tillage would increase significantly, while lower-yielding land would be converted to pasture. However, in a similar study, Stonehouse and Bohl show that meaningful levels of soil erosion prevention via taxation are difficult to achieve and result in significant reductions in net returns.

Beyond the confines of conservation tillage, reviews of new conservation schemes in Europe can provide some insight into the effect of policy on conservation behaviour among farmers. These schemes have developed through a gradual conversion of the European Union's extensive subsidy regime from supporting production to supporting environmental practices such as set-aside. Based on surveying in Scotland, Wynn *et al.* show that compensation alone does not ensure conservation programme success as a lack of awareness of such programmes can limit participation. Once aware, farmers were more likely to participate, as long as there was a good fit with the farm situation and the costs of compliance were low. Compliance costs are often an obstacle to adoption. Even with full compensation for foregone agricultural income resulting from participation, administrative or transaction costs equal to just 5 percent of total compensation can inhibit farmer participation. This evidence from Europe suggests that financial support alone is not sufficient to encourage the adoption of CA-type practices. It is necessary to combine such support with other efforts directed at the specific needs of farm operations.

How Policy Can Enhance the Adoption of Conservation Agriculture

Given the perceived environmental impacts over the past half-century, some have argued that the decoupling of agricultural support from production decisions would represent the most effective means by which governments could alleviate environmental degradation. There is debate concerning the means, both direct and indirect, by which governments can promote conservation in agriculture effectively.

In promoting CA, a key concern for policy-makers is whether CA provides a positive or negative net return to potential adopters. Once this uncertainty is rectified, Uri recommends:

— education and technical assistance where conservation is profitable but the farmer is not aware of the technology or its profitability, or does not have the skills to implement it;
— financial assistance where conservation is not profitable to the individual farmer but would provide substantial public benefits;
— long-term research and development;
— land retirement; and
— regulation and taxes where conservation behaviour is required of all farmers, or for those participating in related income support programmes.

With respect to the first approach, McNairn and Mitchell argue that encouraging the adoption of conservation practices requires assurance of long-term benefits from adoption; unambiguous, easily understood and accurate information; and the promotion of multiple economic and non-economic benefits. Education plays a key role in motivating adoption and requires tailored, credible, and appropriate information and experience that is communicated through the proper channels. Extension services to provide information and assistance can be highly effective, especially in the case of new or emerging technologies, although public agents need not be the exclusive providers of such services.

Financial assistance for the adoption of various conservation practices is well established in Europe and, to a lesser degree, North America. Assistance can take a variety of forms, such as tax credits on equipment, machine rentals, cost-sharing programmes and direct subsidies. Assistance is most suitable to help overcome significant initial investments and transition costs, and in cases where adoption is unprofitable from the individual farm perspective.

A less interventionist policy approach might focus on research and development to enhance the benefits of CA adoption by improving performance or reducing costs. This approach relies on voluntary adoption and aims to increase the odds of this occurring by making the practice more attractive. However, research and development is a long-term policy strategy with an uncertain probability of success.

Land retirement is only suitable in instances where soil erosion concerns are so significant as to warrant conversion to permanent cover crops. Typically, this approach requires significant public financing to compensate farmers, and it is infeasible in areas highly dependent on a limited land base for the production of foodstuffs.

Finally, although tried in some locations, regulating soil erosion limits is not a common approach. This situation probably arises from political awkwardness and onerous enforcement/compliance demands. This is especially so where meeting a soil loss regulation through use of no-till results in significant declines in net returns. A more common regulatory approach involves cross-compliance measures whereby eligibility for a support programme depends on the adoption of certain conservation practices. Because compliance is by choice, programme implementation is liable to be more politically feasible and economically efficient. With respect to the use of taxes on soil erosion, it is possible to induce CA adoption and even pasture conversion. However, meaningful levels of soil conservation involve significant revenue losses. Hence, although possible, taxation is politically infeasible.

The inconclusive nature of empirical studies, and obvious site-specific nature of many results, suggests that a universal approach is not possible. In order to accommodate differences between farms, farmers and economic circumstances, a targeted policy approach may be preferable. In other words, policy mechanisms such as grants or extension services could be geared to the particulars of a location or, preferably, to individual farmers and their farm operations. While a targeted policy approach places a heavy administrative burden on policy-makers, it could achieve greater efficiencies than a more uniform approach, and may represent the most effective means of encouraging CA adoption.

Although a targeted policy approach may be most appropriate for the design of programmes directly promoting CA, there are some alternative policy prescriptions that may be more universally applicable. For example, Isham points out that parallel investments in social capital may be necessary to create a sufficiently enabling environment for the adoption of desirable project activities, and this may apply strongly in the case of CA. Some authors argue that social capital is a product of a learning process. Fostering discussions about the community and seeking consensus decision making can help achieve such learning. A key question is whether governments can foster social capital, as top-down efforts may not be able to promote bottom-up social capital. However, Sobels *et al.*, suggest this is not so, citing Landcare in Australia as an example of successful government support contributing to social capital. Indeed, to a certain degree, the success of Ontario's Environmental Farm Plan programme is ascribable to farmer pride and interest in 'doing the right thing'. Both pride and peer pressure may be important forms of motivation for CA adoption, and government policies may be able to contribute on this front.

Implications for Economic and Policy Analysis

Specialized policy and economic analyses are prerequisites for the appropriate design and correct targeting of CA policies. Policy analysts and economists interested in CA can make use of numerous new techniques and ways of thinking. Sustainability indicators are one example. These capture changes in farming practices that alter the sustainability of the farming system in some quantifiable way that conventional analysis may fail to capture. Therefore, sustainability indicators help describe the evolution of soil productivity over time or present its status in terms that better contrast conditions under CA and conventional management. Sustainability indicators are applicable at the local farming-systems level, at intermediate levels such as the community or region, or at higher levels.

At the village and farm level, sustainability indicators assess the sustainability of specific farming systems and, by inference, the sustainability of soil tillage within a given farming system.

At the macroeconomic level, the system of national accounts has integrated soil degradation through formal green accounting initiatives such as the United Nations System of Integrated Environmental and Economic Accounting. In keeping with standard national accounting practice, green accounting measures disinvestment or investment in soil natural capital and then adjusts NNP/GNP accordingly. Other national indicator approaches include the World Bank's calculations of genuine savings rates. These adjust net domestic savings for changes in the value of resource stocks and pollution damages while the Pearce-Atkinson indicator incorporates elements of the genuine savings idea. Indicators such as this can convey the message powerfully to decision-makers that soil degradation is resulting in a loss in national wealth, and so encourage greater efforts to promote more sustainable practices such as CA.

Analysts who have to assess the attractiveness of projects involving CA or competing farming practices can adopt a number of measures. Such efforts are important because some of the benefits of adopting CA do not show up in conventional cost-benefit type analyses, or in comparisons of CA and alternative practices in narrowly-defined financial terms.

Non-market valuation techniques

It is common practice to use non-market valuation techniques to incorporate the benefits and costs of farming practices that are not priced in markets. Examples include downstream siltation from soil erosion, or loss of organic fertilizer where dung is used as a fuel instead of on farm fields. The valuation practices most appropriate to comparisons of CA and conventional farming practices include replacement cost, changes in productivity, direct and indirect substitute approaches, preventive or mitigative expenditures, and hypothetical or constructed market techniques.

Depletion of soil as natural capital

Economic analyses at the project level can incorporate the depletion of soil as a form of natural capital under conventional tillage practices, so enabling fairer comparisons with CA. This depletion constitutes a cost of non-sustainable cropping in addition to normal production costs. It is a user cost as it yields short-term gains at the expense of future income. Omitting user costs results in an overstatement of the net economic benefits of current cropping practices that deplete soils. Several techniques are available to calculate the user cost of depleting natural resource stocks. Two common approaches are the net price method and the marginal user cost method.

Whole-farm budgeting

Proper environmental analysis requires the assessment of

changes in environmental conditions in terms of the full range of behavioural responses that occur. When farmers adopt CA, numerous ancillary changes can be expected, such as crop switching, changes in pest control measures, shifts in cropping duties for household members (by gender), etc. For this reason, comparative analyses of CA and alternative practices should adopt a whole farm approach to capture the full range of these behavioural changes. Diebel *et al.* argue that analysis of individual practices in isolation can even provide misleading results when certain factors combine synergistically to raise barriers to adoption that are not otherwise evident.

Alternative project evaluation techniques

While project work makes universal use of cost-benefit analysis, other project evaluation techniques hold promise for the appraisal of CA projects or technologies. These include multi-criteria analysis (MCA), cost-effectiveness analysis, decision analysis, environmental impact assessment and participatory methods. MCA recognizes that government decision-makers and smallholders have many objectives in mind when deciding about agricultural project viability and on-farm management practices, respectively; more than a cost-benefit analysis alone can capture. In addition, various trade-off techniques, such as trade-off curves or more sophisticated analytical techniques, can help assess the trade-offs amongst competing objectives. For example, Van Kooten *et al.* use such a method to examine the trade-offs between net returns and stewardship motivations amongst farmers in Saskatchewan, Canada, in adopting soil conservation practices.

The benefits of CA range from supporting basic agricultural production and meeting food security needs in a sustainable manner, to supporting globally important terrestrial and soil-based biodiversity, culminating in carbon sequestration. This review of current thinking about these benefits suggests that the expansion of CA across many different agro-ecological zones makes good sense from a social perspective.

However, the financial profitability of CA is uncertain. Although there appears to be a small cost advantage over conventional practice in general terms, results are liable to fluctuate widely from site to site, with many studies showing CA as less profitable. There are also differences in analysing cases in developed versus developing countries, with tropical hilly examples from the latter group demonstrating distinct advantages for CA because of its more comprehensive approach and better agroclimatic conditions. In contrast, caution is warranted in temperate areas, as the CA approach promoted is less intensive and any cost advantage is likely to be insufficient for bringing about the levels of adoption and diffusion justified from a social perspective. In part, this situation occurs because farmers cannot capture the many national and global benefits from CA.

Given this divergence between private and social interests, interventions promoting more sustainable farming techniques are justifiable in a social sense, and at both the national and international levels. However, CA is not the only soil and water conservation technique that can generate the benefits cited above. Thus, it is necessary to situate CA within a broader range of alternatives to conventional farming practices. Encouragingly, CA is representative of a group of improved agronomic practices that are generally more profitable than competing soil and water conservation technologies that are more structural or purely vegetative in nature.

If CA-type approaches are preferable to the alternatives, then providing monetary compensation to induce adoption might seem an appropriate policy response. However, such an exercise is unlikely to bridge the gap between socially desirable levels of adoption and actual farmer behaviour on its own. Other factors affect adoption as well. For example, numerous such influences are statistically significant in models that attempt to explain actual adoption behaviour (as opposed to general discussions lacking empirical support). These other factors stem from different farmer management objectives, stewardship motives and fundamental barriers or constraints that inhibit a response to profit signals. In some cases, it is the collective rather than the private dimension that is critical to adoption success. There appears to be a correlation between higher levels of social capital and success in these situations. Thus, promoting CA must start with the identification of all factors that impeded adoption and not just a lack of financial net returns.

Policy has also been an important determinant in explaining past CA adoption or non-adoption. Policy stances have sometimes been weak and ineffective in promoting CA. Much of the successful diffusion of the technology has occurred because of support from private corporations, the formation and operation of farmers' groups and other non-governmental pathways. Moreover, conflicting policies have often operated at cross-purposes, encouraging and discouraging CA at the same time. Despite these shortfalls, examples of successful policy measures include green decoupling programmes in Europe and farmland stewardship programmes such as Landcare in Australia.

The above analysis contains implications for policy-makers. On the one hand, an assumption that CA will spread on its own in some desirable fashion is not appropriate. On the other hand, a uniform policy prescription to fit many locations is not realistic either, whether it consists of direct interventions or more indirect incentives stemming from research and development, or some mix of both. Designing successful policies to promote CA is likely to start with a thorough understanding of farm-level conditions. This understanding needs to include management objectives, attitudes to risk, willingness to make trade-offs between stewardship and profits. The next step is the careful design of location-sensitive programmes that draw on a range of policy tools. Flexibility is liable to be a key element in policy design to promote CA.

One area where policies of a more uniform nature might be useful is in the development of social capital and the promotion of the precursor conditions for collective action. For example, the social capital benefits of group extension approaches probably are under-appreciated. Given the demonstrated importance of farmers' groups and information dissemination in the successful diffusion of CA, efforts to strengthen the enabling conditions that foster these activities can pay large dividends.

In devising appropriate policies relating to CA and, more generally, sustainable agriculture, there is a need for improved policy analysis and information for decision making. Developing sustainability indicators that can more clearly show the benefits of CA over its alternatives is one step. Similar improvements are achievable at the economic-analysis level. For example, incorporating the depletion of natural capital in studies of conventional farming practices can help evidence the limitations of these techniques. Ultimately, a whole-farm systems approach may be the most appropriate basis for financial analyses of CA, as this can capture the full range of responses that farmers make when choosing to adopt a new technology such as CA. Moreover, it can incorporate the many options available to farmers in making such choices, something which is not possible in a simplistic comparison of conventional tillage and CA.

Bibliography

Ahiduzzaman, M. 2007. Rice husk energy technologies in Bangladesh. *Agricultural Engineering International: the CIGR Ejournal*, Invited Overview No. 1, Vol. IX.

Aulakh, K.S. 2005. *Punjab Agricultural University, accomplishments and challenges.* Ludhiana. 23 pp.

Bayer, D. & Hill, J. 1993. Weeds. *In* M. Flint, ed. *Integrated pest management for rice*, Vol. 3280, p. 32–55. University of California and Division of Agricultural and Natural Resources.

Belder, P. 2005. *Water saving in lowland rice production: an experimental and modelling study.* Wageningen University, Netherlands. (Ph.D. thesis with English and Dutch summaries)

Bot, A. & Benites, J. 2005. *The importance of soil organic matter, key to drought-resistant soil and sustained food production.* FAO Soils Bulletin No. 80, Rome, FAO.

Busi, R., Vidotto, F., Tabacchi, M. & Ferrero, A. 2002. A preliminary study on propanil-resistant Echinochloa crus-galli in north-west Italy rice fields: *Proc. VII Congress of European Society for Agronomy*. Cordoba, Spain, 15–18 July 2002.

Derpsch, R. 2005. The extent of conservation agriculture adoption worldwide: implications and impact. *Proc. Third World Congress on Conservation Agriculture*, Nairobi, Kenya, 3–7 Oct. 2005. Harare, ACT.

Doets, C.E.M., Best, G. & Friedrich, T. 2000. *Energy and conservation agriculture*. Occasional paper, FAO SDR Energy Programme, Rome.

FAO. 2000. *Global assessment of soil degradation (GLASOD)*. FAO. 2006. *World agriculture: towards 2030/2050*. Interim report, Rome. FAO. 2005. *The state of food and agriculture 2005*. Rome.

Ferrero, A. 1995. Prediction of *Heteranthera reniformis* competition with flooded rice using day-degrees. *Weed Research*, 36: 197–201.

Jat, M.L., Chandna, P., Gupta, R., Sharma, S.K. & Gill, M.A. 2006. *Laser land levelling: a precursor technology for resource conservation.*

Rice-Wheat Consortium Technical Bulletin Series No. 7. New Delhi, India, Rice-Wheat Consortium for the Indo-Gangetic Plains. 48 pp.

Kerr, P. 2001. *Controlled traffic farming at the farm level.* GRDC Research Update, Finley, NSW, Australia.

Met Office. 2005. *Climate change, rivers and rainfall.* Recent research on climate change science from the Hadley Centre, Exeter, UK.

Michigan, American Society of Agricultural Engineers. Kartha, S. & Larson, E.D. 2000. *Bioenergy primer: modernised biomass energy for sustainable development.* New York, UNDP/BDP Energy and Environment Group.

Miyamoto, K. 1997. *Renewable biological systems for alternative sustainable energy production.* FAO Agricultural Services Bulletin No. 128. Rome.

PAU. 2006. *Farm equipment research and development activities.* Department of farm power and machinery information brochure, Ludhiana, India, Punjab Agricultural University.

Ragab, R. & Prudhomme, Ch. 2002. Climate change and water resources management in arid and semi arid regions: prospective and challenges for the 21st century. *Biosystems Engineering,* 81(1): 3–34.

Reicosky, D.C. 2001. Conservation agriculture: global environmental benefits of soil carbon management. *In* L. Garcia Torres, J. Benites & A. Martinez-Vilela, eds. *Conservation agriculture, a worldwide challenge,* Vol. I, p. 3–12. Rome, FAO and Brussles, ECAF.

Robinson, A.P., Hollingdale, A.C.; & Reupke, P. 1993. *The construction and operation of a rice husk burner.* Rural Technology Guide 18. Chatham, UK, Natural Resources Institute.

Saturnino, H.M. & Landers, J.N. 2002. *The environment and zero tillage.* Brasilia, Brazil, APDC-FAO. 144 pp.

Scott, D.S. & Legge, R.L. 1999. Biomass feedstocks. Biocrude oil. *In* O. Kitani, ed. *CIGR handbook of agricultural engineering.* Vol V. *Energy and biomass engineering.* St Joseph, Michigan, American Society of Agricultural Engineers.

Shaxon, T.F. & Barber, R.G. 2003. *Optimizing soil moisture for plant production—the significance of soil porosity.* FAO Soils Bulletin No. 79, Rome, FAO.

Solomon, B.D, Barnes, J.R. & Halvorsen, K.E. 2007. Grain and cellulosic ethanol: history, economics, and energy policy. *Biomass and Bioenergy,* 31(6): 416–425.

Stout, B.A. 1989. *Handbook of energy for world agriculture.* London, Elsevier Applied Science.